UNDERSTANDING AND DOING MATH

CIRCLE 1

What are Math Objects?
Illustrated with Numbers

Boris Čulina

Understanding
Zagreb, 2021

UNDERSTANDING AND DOING MATH
CIRCLE ONE

What are Math Objects?
Illustrated with Numbers

Boris Čulina
Zagreb, 2021.

Front Cover:
Boris Čulina
Bruno Kokot
Steve Grundy

Cartoons:
Boris Čulina
Ivan Filipović

English Translation
Marina Manucci
Patrick Burton

Printed by
Amazon print-on-demand

Publisher
Understanding

ISBN: 978-953-8387-00-5 (Series)
ISBN: 978-953-8387-01-2 (Part 1)

CIP record available at
https://www.nsk.hr/en
in NSK Catalogue under number 001097097.

Contents

Preface to the Circles Series

Learning is much more similar to biological growth than to manufacture, where component parts are first produced, then fitted together.

W. Servais, T. Varga: *Teaching School Mathematics,*
A UNESCO Source Book, 1971.

During the twentieth century, mathematics experienced a real explosion of content and methods. Its modern core, however, consists of only a few principles. Acquiring these principles, understanding them, and learning the art of combining them in modelling more complex situations forms the mathematical minimum that I would like to present in these books (the circles). My goal is to present these principles in a simple – yet not simplified – manner, so that the reader actually acquires them, and does not just have the impression that she or he has acquired them. My goal is to present mathematics in a way that helps the reader to develop the mathematical dimension of his (her) humanity and does not just show him (her) the human dimension of mathematics. All the books present the same principles of mathematics, but at different levels of abstraction and complexity. The basic philosophy of the Circles Series is that *the acquisition of mathematics is not like constructing a building brick by brick, but more like the growth of a living organism in which the same principles that are present in the very beginning develop in each phase of growth.* These books have been conceived so that you can study them one after the other until you achieve a level of understanding with which you will be satisfied: a level that will allow you to acquire the appropriate mathematical dimension for your personal activities. You do not even have to start with the first book, but rather with the book you consider to be the right starting point to fill the gaps in your mathematical knowledge. The books are designed for self-study, and each book can be studied independently of the others. They cover the range from elementary mathematics to the mathematical foundation for any non-mathematical profession. The first book assumes that there has already been a "zero[th] book" – that the reader have acquired certain calculating skills and a certain dissatisfaction with her (his) understanding of what she (he) have been calculating. The first two books are intended primarily for teenagers. They enable them to gain confidence in their mastery of numbers, to gain a broader understanding of mathematics through the example of numbers and, eventually, to get rid of the frustrations that make it difficult for them to learn mathematics. This does not mean that one of the older readers cannot go back to their teenage years and make up for the missed mathematical knowledge.

Furthermore, each book is not a world in itself. Each one contains links to its own web page, `https://understandingmath.academy/math-circles/`, where auxiliary content can be found, as well as to `https://en.wikipedia.org/wiki/Portal:Mathematics` which is truly an enormous source of high-quality mathematical knowledge. If you have additional interest in particular topics, the links will guide you to the appropriate web pages.

Exercises in mathematics play the same role as practical work in mastering practical activities. Without them, it would be impossible to acquire the ability to actually do mathematics. You can know by heart the entire theory of processing wood, but if you do not invest enough time in physically working with wood, you will certainly never become a good carpenter. The problems accompanying each book let you check whether you have correctly acquired both the theory and the skill of its application. Therefore, each chapter is followed by problems and their solutions. They have been classified according to the topics of the chapter, so that when you have studied a certain topic, you can immediately solve the corresponding problems at the end of the chapter.

Although the focus in these books is on comprehension rather than calculation, calculating is an essential component of mathematics. Acquiring mathematical principles also requires adequate elementary calculation skills. Sooner or later, the need for more complex calculations may come along. Fortunately, this is where computers can be of significant help nowadays. An important part of acquiring core mathematical skills is developing the skill of using the computer in mathematical activities. On the one hand, we have powerful mathematics software which require time to learn how to use and are not free of charge. On the other hand, there are a number of simple, free math software, but each of them solves only a certain type of problem. Fortunately, in recent decades a simple and advanced free open-source software called *SageMath* has been developed. *SageMath* is essentially an elegant amalgamation of all the previous big mathematical open-source software and has become a serious competitor to professional mathematics software. We will use *SageMath*'s simplest mode – when we need it, we will just visit the web page `https://sagecell.sagemath.org/` and key in the appropriate commands in the marked space. The *SageMath* tutorial and the *SageMath* examples for each book can be found on the web page `https://understandingmath.academy/sagemath-materials/`.

We will start studying math together with two different students, an *Astonished* guy and an *Uninterested* one. Each of them has one dominant psychological characteristic in the learning process, both of which are present in ourselves in different ratios. My teaching will be supplemented by the *Professor*. He has all the characteristics of a good teacher that I lack.

When we were nearing the end of compiling material for the books, my cartoonist Filipović and I felt that we were missing something important. We finally realised we were minus a female character! What a subconscious male embarrassment! We imagined that she would be a critical and direct down-to-earth punk girl to whom you can't sell everything, and the cartoonist had already started to design her character. Unfortunately, my dear Filipović died and our *DowntoEarth* girl was never realized. I imagine that she is still here, so I ask you to imagine her presence while reading these books. They will definitely be better that way.

Many examples in this book are related to my own life and the environments I have lived in, mainly in Croatia and in Bosnia and Herzegovina. Whether you are from New York or from Madagascar, this does not change a thing. Our concrete lives are the best way to share universal human values.

A brief description of each book's contents can be found below:

UNDERSTANDING AND DOING MATH – CIRCLE 1: What are Math Objects? Illustrated with Numbers. The familiar example of numbers is used to illustrate what mathematical objects are, how we manage them, and how we apply them.

UNDERSTANDING AND DOING MATH – CIRCLE 2: What are Math Tools? Illustrated with Numbers. At a higher level of precision than in Circle 1, various mathematical tools are described, as well as how they are applied in modelling more complex situations.

UNDERSTANDING AND DOING MATH – CIRCLE 3: Mathematics of Geometry and Geometry of Mathematics The goal of this Circle, in addition to the acquisition of geometric knowledge and related mathematics, is primarily for the reader to acquire geometric ideas and their mathematical realization, and to prepare him (her) for learning the geometry of mathematics: how geometric ideas put light into the most abstract parts of mathematics and thus make them more efficient.

UNDERSTANDING AND DOING MATH – CIRCLE 4: Mathematics: From Ideas to Realizations. In addition to acquiring two important and mathematically different parts of mathematics (continuous mathematics of change versus discrete mathematics of graphs), the main purpose of this part is to show the reader, using these examples, how to go from an idea to its mathematical realization.

UNDERSTANDING AND DOING MATH – CIRCLE 5: Mathematical Ideas and Structures. At a higher level of precision than in the previous books, basic mathematical ideas are analysed and mathematical structures for the realisation of these ideas are developed. These structures, together with methods for their combination and application, form the core of modern mathematics.

UNDERSTANDING AND DOING MATH – CIRCLE 6: The Architecture of Mathematics. At a level of precision characteristic of advanced mathematics, the logical structure of modern mathematics, as a distinctive and integral area of human activity, is presented.

I am deeply thankful to Ivan Filipović (cartoons), Marina Manucci (English translation), Patrick Burton (editing English translation), Bruno Kokot (the book covers), Ivica Bilušić (editing pictures) and Sanja Vitaljić (assistance in preparing exercises) for their friendly cooperation in making these books.

Preface to Circle 1

Mathematics seems to endow one with something like a new sense.
Charles Darwin, originator of the theory of evolution (1809–1889)

Who cares for math!

my son

There are four great sciences ... Of these sciences the gate and key is mathematics, which the saints discovered at the beginning of the world.

Roger Bacon, philosopher and scientist (1214–1292)

The good Christian should beware of mathematicians and all those who make empty prophecies. The danger already exists that mathematicians have made a covenant with the devil to darken the spirit and confine man in the bonds of Hell.

Saint Augustine, theologian and philosopher (354–430)

It is not a story of brilliant achievement, heroic deeds, or noble sacrifice. It is a story of blind stumbling and chance discovery, of groping in the dark and refusing to admit the light. It is a story replete with obscurantism and prejudice, of sound judgment often eclipsed by loyalty to tradition, and of reason long-held subservient to custom. In short, it is a human story.

Tobias Dantzig, about the development of the notion of number, in *Number: The Language of Science*, Macmillan, 1930.

Mathematics is neither God's nor the devil's, but rather a human activity. Humans have created it in order to manage the world more easily, just as they invented the compass and the computer. Indeed, it can easily encourage emotions, as evidenced by the quotations above. But is it not the same with football or music? While some glorify mathematics and attribute divine qualities to it, others are horrified by it and think that it is necessary to have a special gift in order to understand it. All these are simply myths. Since mathematics is a human creation, it can be understood, practised and used. Everyone can learn how to play a guitar well enough to play simpler melodies and enjoy composing their own. It is the same with mathematics. *You can understand it without problem, acquire it, and use it.* How far you will go will depend only on your reasons for studying it and how interesting you find it.

How much you need mathematics depends on what you do. In everyday situations it is almost unnecessary. Sometimes you need to add up, subtract, multiply, or divide something – usually natural or decimal numbers. Have you ever had – in an everyday situation – to multiply two fractions or calculate the circumference of a circle? However, everyday life is also becoming more complex. You receive more and more information and it is very important that you can distinguish truth from a lie and the good intentions from attempts at manipulating. In the years to come it seems that this skill will be even essential for the survival of Humanity. Hannah Arendt, in her book *The Origins of Totalitarianism*, writes: "The ideal subject of totalitarian rule is not the convinced Nazi or the convinced Communist, but people for whom the distinction between fact and fiction (i.e. the reality of experience) and the distinction between true and false (i.e. the standards of thought) no longer exist." And this is where thinking and knowledge help a lot. *Mathematics is their important part.*

No matter what you do in your life, it is almost certain that there are situations in which mathematics will come in handy, since it is embedded in almost every human product, whether the mobile phone or knowledge about the nuclear processes occurring on the Sun. Maybe it is mathematics that you studied at school, maybe it is mathematics that you will find in specialised books, or maybe you will have to create it yourself. *The main purpose of math educa-*

xiv

tion is not so much to provide you with ready-made mathematical thoughts as to make you ready to think mathematically.

There is one more reason why (some) people deal with mathematics, and that is beauty. *Independent of any benefit, they are attracted by the harmony of mathematical ideas.* However, this beauty is often related to the applicability of mathematics. If the world seems harmonious to us, then the mathematics that will describe it needs to be just the same. The quest for beauty has more than once delivered great discoveries. As an example one can take the physicist and mathematician James Clerk Maxwell (1831–1879). Studying the already-known laws of electromagnetism, it seemed to him that they would be much "nicer" if he changed one equation "a little". Upon doing this, he obtained the laws on whose basis he concluded something completely unexpected – that there exist electromagnetic waves, and that the light itself is an electromagnetic wave. This discovery is the foundation of advanced information transfer from television to internet.

(The text is from the poem *Notturno* by the great Croatian poet Tin Ujević)

Before we start our joint journey, let us explode another myth: the myth about mathematics as unerring and perfect knowledge. The math we find in books is usually a finished product where everything is nicely arranged, as in a store of porcelain figures. Scared of breaking something, you dare not move. However, mathematics in its emergence is completely different. It is full of trial and error, dead ends, and misconceptions. Simply, it is alive. Only in the end all the failures are pushed "under the carpet", and only successes are shown. Everyone knows about Isaac Newton (1642–1727), the giant of thinking who laid the foundations of modern science. However, only a few know that he considered the discovery of the date of the world's origin as one of his most important discoveries. He calculated that God created the World in 3500 BCE. He made a mistake by only some twenty billion years, if not even more. Learning mathematics is actually the emergence of mathematics in your head. *So, do not be afraid of imagination, trials and errors. You will learn math better with them than without them.*

We will start our journey with numbers. *Numbers are the oldest mathematical idea, but still also the most important one. Not only will we go through the basics of numbers in a way that will give you the confidence to really understand numbers and really know how to apply them, but you will also learn all the essential elements of mathematics through the example of the world of numbers.* The example of numbers will be used to illustrate what mathematical objects are and how they are applied, and what mathematical tools we use in their description and application. These elements are summarised in the *Calm Epilogue* of the book.

Humanity needed millennia to develop the world of numbers and methods for their description and application. While growing up you are expected to pass through this history briefly in a few years of education. On the basis of the experience of the whole of human civilisation and your education, we are now in the position to acquire, in several weeks, knowledge about numbers at a more mature level.

Introduction to Numbers

Number rules the universe.

> Pythagoras, 6th century BCE philosopher

Numbers are the free creation of the human mind.

Richard Dedekind, one of the creators of modern mathematics (1831–1916)

At the very beginning I must say that numbers are strange in one aspect. Where are they? Biology deals mainly with plants and animals. We can see precisely what it studies. History deals with humans and events that are not here any more, but whose existence is easy to imagine. Physics deals, for instance, with atoms. We cannot see them, but we can understand where they are and why we do not see them. But, what is it with numbers? And what about other mathematical objects? Some of you will probably have got acquainted with vectors and matrices, maybe with even more complex mathematical objects, such as differential forms on manifolds or Hermitian operators on infinite-dimensional complex Hilbert spaces. But do they really exist or are they as fictional as creatures from fairy tales, like fairies and dwarfs, trolls and gremlins? Isn't it a little bit odd, after you have spent so much time calculating with numbers, that, probably, nobody has ever told you whether they exist at all, never mind where they actually are. This is one of those "embarrassing" questions that people try to avoid. Because the answer is simple and disturbing: *it is not known whether numbers do exist!* For thousands of years, mathematicians and those who helped (hindered?) them have not found an answer. Or, rather, they have found several answers and are still arguing about which one is the right one. You can see the current situation of the discussion at the website `https://en.wikipedia.o rg/wiki/Philosophy_of_mathematics`. But be careful not to get too bound up in it!

The quotations at the beginning of the chapter express two extreme attitudes. The ancient Greek Pythagoras was convinced that numbers not only exist but are the basis of the entire world: the basic laws of nature are mathematical laws of numbers and they determine everything – from the way the stars will move, to with whom you will fall in love. Maybe it is not clear how the wise Pythagoras could say something so unwise, but history has shown that this blind faith of his in numbers has been very fruitful. However, if Pythagoras was right, where in the world are these numbers and how do we reveal the truths about them? More than two thousand years later, Dedekind's thinking took him in a completely different direction. He considered numbers to be a free human creation. But if we did invent numbers, how do they help us manage in the world?

The contemporary viewpoint is closer to Dedekind: numbers and other mathematical objects do not actually exist; what is more, they resemble the creatures from fairy tales. A good fairy tale touches us deeply and teaches us through its world of imagined characters and events that, for instance, love is more valuable than money or we should not judge someone by their name. Similarly, a good mathematical fairy tale attracts attention by its inner harmony and provides us with a powerful tool to solve problems. *The world of numbers is an imagined world, not exactly arbitrarily imagined, but rather imagined for a particular purpose, its use being to count and measure.* For example, each natural number in the process of counting has a precisely defined role. And it is only this role that is important, and not out of what it is made. The same as how chess pieces may be made of wood or plastic, or even be drawings on the computer screen, but it is their role in the game that is important. A piece is a king, not because it was made of iron or beech, but because it can move one square at a time , and if it is checkmated by another piece, the game is lost. It is the same with numbers. The number 3 can be imagined as three small lines on paper or as three knots on a rope. However, the only important thing is its role in the imagined structure (world) of natural numbers – that it is a number following number 2 and preceding number 4. This is also my view of numbers, as a world imagined with a certain purpose. Such a view of mathematical objects has a certain advantage, since it leaves us the freedom to imagine any worlds, regardless of whether they do actually exist or not. This view of mathematics also proved historically significant. It was giving up on the older philosophy of mathematics (according to which mathematics describes eternal truths about eternal objects) and accepting the new philosophy, in which mathematical worlds are free products of our mind, that liberated human mathematical powers in the 19th century and led to the true flourishing of modern mathematics that still continues today.

> ### Mathematical Worlds
>
> Mathematical worlds are imagined worlds with the purpose of being powerful tools in studying and shaping the world, or to put it more concretely, in solving problems.

Although I will stick to Dedekind's approach to mathematics, you do not have to agree with my opinion, just as many other mathematicians and philosophers have not. If you find it more acceptable to think that numbers do exist in a certain way and that we did not invent their properties but just discovered them, feel free to imagine them like that. Once we have agreed on their properties and method of functioning, this issue will have no further impact on our work. Mathematics is such a powerful and lively human activity that it functions perfectly well even without a final answer to the question of whether mathematical objects exist or not. *When the world of numbers, real or imagined, started to be connected with the real world in the processes of counting and measuring, the human community came to unprecedented knowledge and the possibilities of control.* This started with the ancient Greeks: first of all with Pythagoras. He was probably the first who understood what immense power is given to us by numbers, so maybe it was in this enthusiasm that he exclaimed that numbers manage the world. However, the first to implement this power in practice were the creators of modern science in 17th century: Galileo, Kepler and Newton. They and their successors, by counting, measuring, and thinking, provided the description of nature which is the basis of our modern-day understanding of universe and the technical advances that facilitate (or aggravate) our life.

If, by any chance, Pythagoras and Dedekind were to meet today, although they would not be able to agree on the issue of the real nature of numbers, they would most certainly agree that *the numbers are the basis of our description of the world*. Without numbers we can state that there is a lot or a little of something, but with them we can resolve this much more precisely. For instance, there are exactly 17 books on my shelf and by counting them I can easily find out whether there any are missing. Also, in human language there are only a few words for the common colours (blue, yellow, dark red, etc.). However, with the use of numbers we can say very precisely that, for example, the yellow colour of helium is not just any kind of yellow but rather a yellow of a particularly defined wavelength of 5.88×10^{-7} metres. Not only can we use measuring to determine something more precisely, but also regularities in nature can be expressed precisely by the relations between numbers. For instance, when a steel bar is heated, its elongation Δl is proportional to the existing length of the bar l and the change in temperature Δt ($^{\circ}C$ denotes the degree Celsius – a unit for temperature):

$$\Delta l = 16 \cdot 10^{-6} \, ^{\circ}C^{-1} \cdot l \cdot \Delta t$$

This law should be considered in the case of all steel structures, ranging from railway lines to bridges. It can be used to calculate exactly the elongation of the steel elements, and hence design the structures so that such elongation does not disturb their stability or functionality.

These are some almost randomly selected examples that show how numbers are important and why good knowledge of them is necessary. They also illustrate the essential characteristic of all good mathematical ideas – *they are a powerful tool for investigating and designing the world*.

Numbers are just one of the realisations of the general idea of measurement. In this first circle we will get to know also a different realisation of this idea, the coordinate system. There are also some other very important realisations of the idea of measurement which I will only mention in the *Inspiring End* of the book, and some of them will be studied in subsequent Circles.

Measure what is measurable, and make measurable what is not so.

Galileo Galilei, scientist (1564–1642)

1

Natural Numbers (and some other unnatural phenomena)

In this chapter we will recollect our knowledge about natural numbers:

- how we imagine them and what their purpose is;

- how we write them down;

- why and how we compare, add, subtract, multiply, divide and raise them to the power;

- what the properties of these operations are.

We will see why in mathematics when somebody uses a word, you have the right to ask for its definition, and when somebody claims something, you have the right to ask for a proof. Next, we will talk about the difference between "content-wise" and "formal" in mathematics, why understanding is important, why formal procedures are important, what it means to calculate and simplify the descriptions of numbers, and how this is performed. We will also see why in mathematics all this is done in an unusual language that is called mathematical language. Finally, we will pay attention to the background mathematical objects that are hidden in every mathematical consideration – sets, relations and functions.

1.1　Counting

Astonished is very proud of his CD collection. However, he has noticed that his friend Uninterested very often "borrows" a CD or two. How can he easily observe when this happens? Fortunately, a long, long time ago people invented numbers and counting in order to be able to control the quantity of items in a set. Astonished has known this from childhood and so he can easily count his beloved CDs:

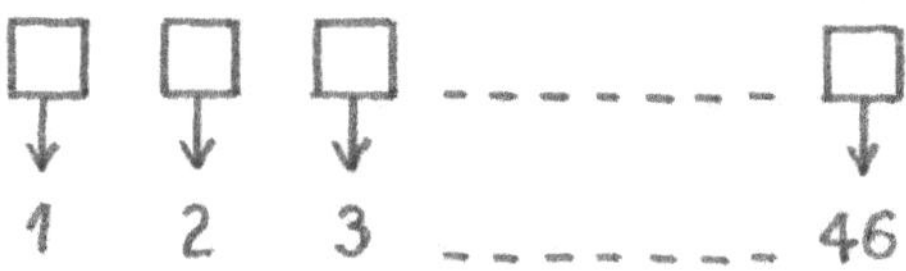

Thus, he has 46 CDs. In the counting process Astonished assigned numbers to his CDs. *The numbers are nothing but imagined objects with which we count.* In order to fulfil their purpose it is important only that they meet the following two (see, counting again!) conditions:

1. *There must be the first number from which we start to count.* It is designated by the symbol '1' (read: *one*), and we assign it to the first item of the set that is being counted:

2. *After each number, there must be the next new number which we will use, as necessary, to continue counting.* It will be assigned to the next selected item from the set. The next number following 1 is denoted as 2 (*two*), the next after 2 is denoted as 3 (*three*) etc.:

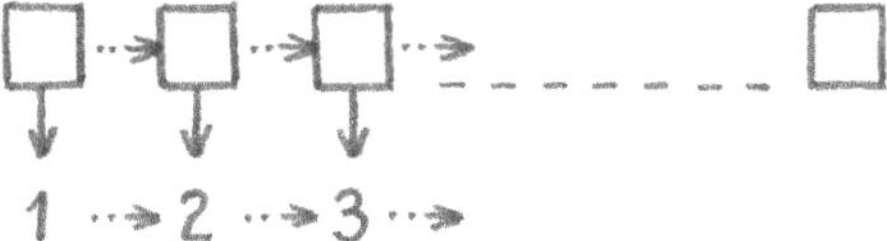

We continue counting as long as there are non-counted items in the set. The number assigned to the last selected item is taken as the measure of the quantity of items in the set and we call it the **number of elements in the set**:

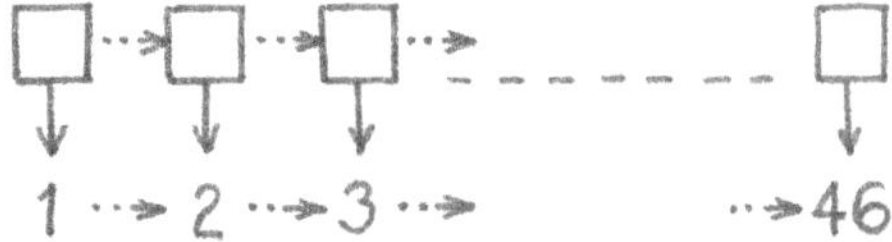

Thus, Astonished has 46 CDs. If he gets a new CD as a gift, the counting will stop at the next number, so he will have 47 CDs. If, however, his friend Uninterested takes one CD, the counting will stop at the previous number 45 and he will notice it.

> **Numbers as a Human Cognitive Tool**
>
> Numbers are imagined objects that make it possible for us to control reality, and this is something that neither ducks nor rabbits possess.

In this way we could count the elements of any set except one. If we want to sum up how many audio CDs Astonished's grandma has got, we will probably find out that she has none. However, how do we express this in numbers, when we cannot even start counting? It is simple: we will invent a new number. This new number will be designated by the symbol '0' (read **zero**) and we will place it in front of all the other numbers. The number 0 denotes the state of counting before we select any item from the set that is being counted. If a set has no elements, then counting stops at the very beginning, at the number 0. Thus, we can now count even grandma's CDs. Sorry, grandma, but you have got zero CDs.

Natural Numbers

The numbers that are used to measure how many objects are in a set (to count them) are called **natural numbers**. We imagine them as a sequence of items used for counting: 0, 1, 2, 3, ...

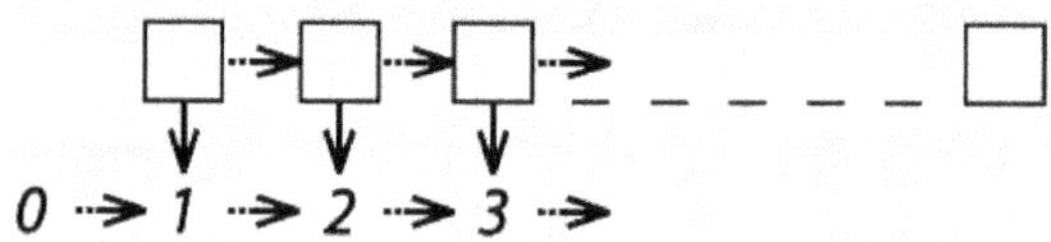

The number 0 is the first number from which we start to count. After each number there must be the next new number which we will use, as necessary, to continue counting. This sequence has not got its last member, but rather there is another after each member – this ensures it is possible to continue counting.

The set of all natural numbers is denoted $\mathbb{N}$.

You can read more about counting on the web page `https://en.wikipedia.org/wiki/Counting`.

1.2　Decimal Notation

The number 0 has proved to be useful. In fact, extremely useful. It enables a simple way of notating numbers – one that we have become so used to, it might be difficult to fully grasp how important it really is. If Astonished has 46 CDs, it means that he has 4 groups, each of 10 CDs, and 6 CDs more:

As we count, we group, at the same time, ten smaller units into a single bigger one – ones into tens, tens into hundreds, and so on. Therefore, we need only the ten symbols for the first ten numbers, and these are:

0, 1, 2, 3, 4, 5, 6, 7, 8, and 9.

By adding another one we get 1 ten and 0 ones, and 1 ten and 1 one, etc., and this we simply record as 10, 11, etc.:

0, 1, 2, 3, 4, 5, 6, 7, 8, 9, 10, 11,..., 21,..., 176,..., 406,..., 3258,...

If, for instance, in a prize competition *Buy a Hundred Pairs of Socks and Win a Prize!!* Astonished got 406 CDs, he would immediately know that it was a lot.

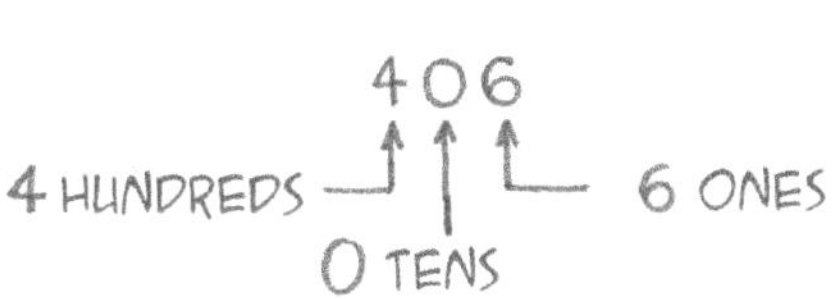

> ### Decimal Notation
>
> Each digit in the notation of a number shows by its position what it counts (ones, tens...), and by its value how many there are. Such notation is called **positional decimal notation of numbers**, and it is made possible precisely by the number 0.

Decimal notation is not only a simple notation of numbers; it also makes it possible for us, as we will see later, to solve many significant issues by simple calculation. Of course, we can make another grouping: let us say 12 lower units into a bigger one (duodecimal system), or, like the computer, two (binary system). Throughout history, the decimal system has prevailed because of the total of ten fingers on human hands.

You can find out more about the decimal system at `https://en.wikip edia.org/wiki/Hindu%E2%80%93Arabic_numeral_system`. You can learn more about the glorious history of "nothing" (zero) at `https://en.wikiped ia.org/wiki/0_(number)`

1.3　Comparison

How will we determine who has more CDs: Astonished or Uninterested? If we did not use numbers, we could do this by combining into pairs, one at a time, Astonished's CDs and Uninterested's CDs:

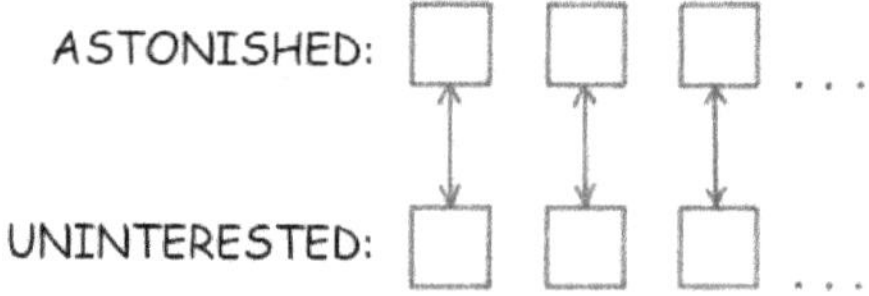

If all their CDs combined like this and there were none left over, we would say that the sets of CDs of Astonished and Uninterested had **equal elements**. If, however, after the combining into pairs was finished, there were still CDs left in one of the sets, then we would say that this set had **more elements**, and the other one **fewer elements**.

Since we imagined the numbers as measures of the quantity of items in a set, we describe the comparison of sets with numbers. Comparison of numbers has been imagined precisely so as to match the comparison of the counted sets.

Comparison of Natural Numbers

Let a be the number of elements in a set A, and b the number of elements in a set B. We say that **number a is lesser (smaller) than number b** and we write $\boldsymbol{a < b}$, if set A has fewer elements than set B. If, however, set A has more elements than set B, then we say that **number a is greater (bigger) than number b** and we write $\boldsymbol{a > b}$. Of course, if A and B have equal elements then the enumeration of each set will stop at the same number: $\boldsymbol{a = b}$.

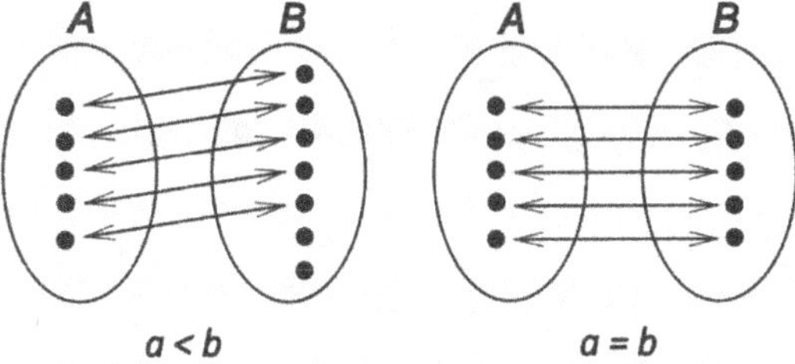

Looking at the sequence of numbers, the lesser number is the one which occurs earlier in the sequence:

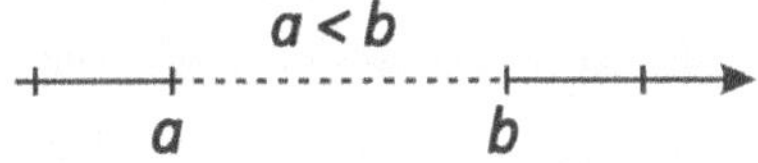

With this description we reduced the comparison of sets to the comparison of numbers assigned to these sets. Let us mention that with "$a \neq b$" we note that numbers a and b are not equal.

But how do we determine which of the two numbers is the bigger one? Simply – by examining their notations! If Astonished has 46 CDs, and Uninterested has 19, Astonished has more groups of 10 CDs (Astonished has 4, and Uninterested 1), and therefore he has more CDs, in spite of the fact that Uninterested has more left over from the groups of 10 (Astonished 6, and Uninterested 9): $46 > 19$. By this rule we can simply determine which number is bigger.

Comparing Natural Numbers

Looking from left to right, we are searching for the first position in the decimal notation of numbers in which the digits differ (if necessary, we can imagine that there are zeroes on the left of the first non-zero digit). The number that has a bigger digit at this position is bigger, and the other one is smaller. If, however, the digits at all positions are the same, then it is the same number.

Such a rule we call **formal**, since it speaks about notations of numbers, and not about numbers nor the items that are being counted. Its real power lies in the fact that it is easily applicable even when huge sets are in question. This is an example of yet another essential characteristic of mathematics, and of other sciences as well:

Content-wise, Formal and the Importance of Good Notation in Mathematics

It is very important to have a good notation in which the content-wise relations (relations between objects that we are describing) are transferred into simple formal relations (relations between linguistic descriptions of these objects). This allows us clearer and more efficient management, especially in more complicated situations. Moreover, the formal part can be turned over to the computer to perform instead of us.

Example 1.3.1. Let us imagine that Astonished and Uninterested have made a fortune (by applying mathematics). Over their entire lives, they have been steadily gaining wealth so that eventually they could fulfil their

dreams: to buy as many CDs as possible. Astonished has bought
249 753 803 245, and Uninterested 249 753 802 345. Who has more CDs?

Solution. If we wanted to determine who has more CDs by direct comparison (content-wise), our lives would be too short. But our formal rule functions just as well as in the case of small numbers. Comparing the digits from
left to right, we will easily determine that the first different digits will occur
at the position of thousands:

$$2\ 4\ 9\ 7\ 5\ 3\ 8\ 0\ 3\ 2\ 4\ 5$$
$$2\ 4\ 9\ 7\ 5\ 3\ 8\ 0\ 2\ 3\ 4\ 5$$
$$\uparrow$$

Since in the notation of the first number at the position of thousands there
is the digit 3, and in the notation of the second number there is the digit 2,
and since 3 is bigger than 2, the first number is bigger. Thus, Astonished
has more CDs. $\square$

Example 1.3.2. If there are 32 pupils in one class, and if the first letters
of their surnames include all of the letters of the Croatian alphabet, can we
claim (without looking at the class register) that the surnames of at least
two pupils start with the same letter? Note: the Croatian alphabet consists
of 30 letters – apart from most of the letters of the Latin alphabet, it also has
the letters *č,ć, dž, đ, lj, nj, š* and *ž* (some letters have two symbols), which
indicate very lovely sounds (see https://www.learncroatian.eu/blog/th
e-croatian-letters).

Solution. Let us assume it is not as above, but that all the surnames start
with different letters. Then the alphabet would have to have at least 32
letters, instead of 30. So, some pupils must have the same first letter in
their surnames. $\square$

The argumentation in the above problem is often used in both mathematics and detective stories, and it is called **proof by contradiction**. When we
want to prove by contradiction that a certain statement A is true, we start
from the assumption that it is not true: that it is false. If from this assumption we obtain a contradiction (that something both is and is not, which is
impossible), then we conclude that the statement A cannot be false. So, it
is true. Why is such an inference correct? And what is correct inference
anyway and how do we infer correctly? These issues will be discussed in
Circle 4. For the moment, it is sufficient to say that we use inference, which
we use also in everyday life, and which we will gradually "discipline" during
our study.

1.4 Addition

If Astonished and Uninterested joined their CD collections into a common collection, how many CDs would they have then? We could count them all anew, but it is better to develop a little bit of theory that will help us in such situations. We will use the already-known data that Astonished has 46 CDs and Uninterested has 19. The total number of CDs will certainly depend on these numbers and nothing else. That is why we can describe (identify) it with these numbers (46 and 19) and the combination method that we mark by the symbol '+'). The total number will be denoted by '$46 + 19$'.

Addition of Natural Numbers

Let A and B be sets that have no common elements. Let a be the number of elements in set A, and b the number of elements in set B. Then the number of elements in both sets together is called the **sum of the numbers** a and b and we write $\boldsymbol{a + b}$ (read a **plus** b). The symbol '+' ("*plus*") is the mark of the operation called addition that assigns to two numbers a third number – their sum.

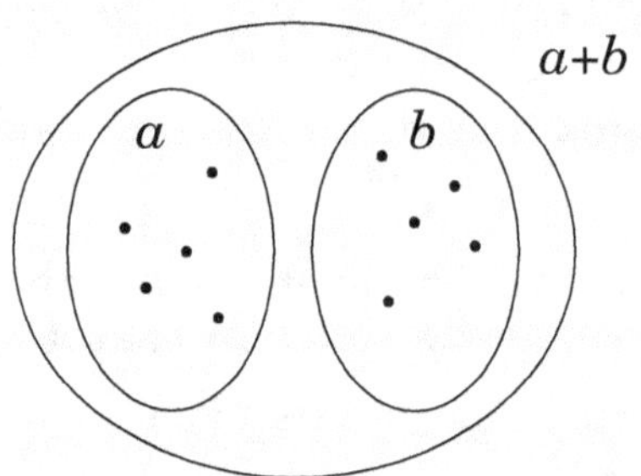

In the sequence of numbers we will obtain the number $\boldsymbol{a + b}$ by starting from number a and counting for b units in the direction of larger numbers:

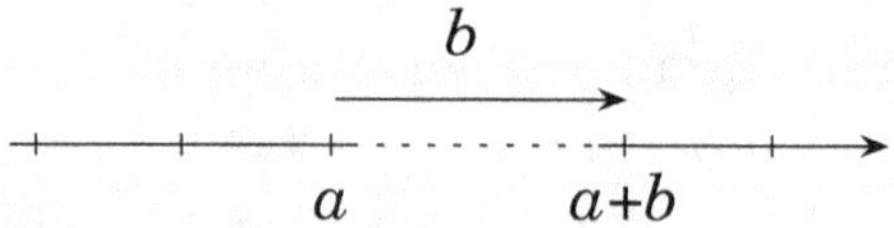

Using the operation of addition we can describe how many CDs Astonished and Uninterested will have together: there will be $46 + 19$ CDs. But how do we find this number? Again, simply – by using the decimal notation in the way we know from primary school:

$$
\begin{array}{r}
46 \\
+\,19 \\
\hline
65
\end{array}
$$

We added Astonished's 6 CDs and Uninterested's 9 CDs and obtained 15 CDs. We noted down 5 ones and continued adding 1 ten to the other tens, so that we have $1 + 4 + 1 = 6$ tens. The procedure, as well as the number comparison procedure, is also of a formal nature, since it consists of actions not on numbers but rather on number notations. In the case of decimal notation, these actions are particularly simple, unlike, for instance, the Roman notation of numbers. In decimal notation, the entire procedure of addition is reduced to addition of the numbers from 0 to 9, and the results of such additions can be recorded in a table.

The Importance of Formal Procedures

Because of the formal nature of the procedure, anyone can learn it and correctly find, for example, the sum of any two numbers, not having to know what numbers are and what the sum of the two numbers means. This is illustrated precisely by calculators and computers, which calculate perfectly well despite not knowing the purpose behind their calculations. However, the essential property of a formal procedure is its correctness – that it gives the right answer. And we can achieve it only by content-wise consideration.

The Latin translation of the work of the Persian mathematician Muhammad ibn Muse al-Khwarizmi (780—850) introduced into Europe the Indian digits (today we call them Arabic digits), decimal notation and simple formal rules of calculation. The very word 'algorithm' (routine procedure) originates from Algoritmi, the Latinised version of the mathematician's name.

Example 1.4.1. On the occasion of the third day of autumn, a party was organised and 36 politicians and 16 CEOs of city companies were invited. When he heard this, the chef made special efforts and ordered $36 + 16 = 52$ meals to be prepared. Although everyone invited came, there were only 42 of them. The unfortunate chef insisted that he had checked his procedure of adding but did not find his mistake. Where did he go wrong?

Solution. The chef's addition procedure is correct:

However, he applied addition incorrectly. It is only when the sets have no common elements that the total number of their elements is equal to the sum of the elements of the single sets. Obviously, this was not the case here. Since the sum is greater by ten than the actual number of guests, it means that ten of them were counted twice. These are those who are both politicians and CEOs of city companies.

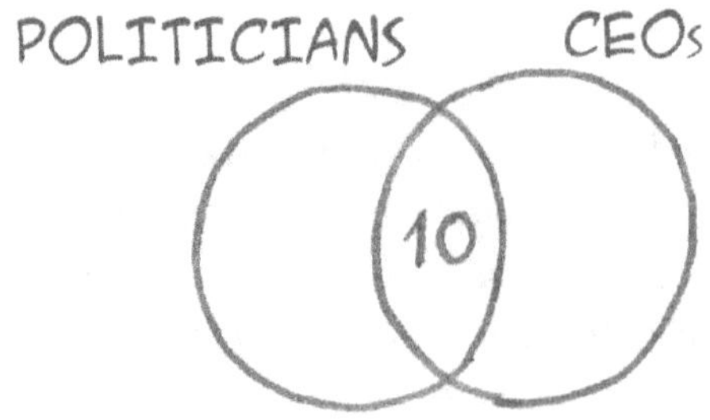

> ### Understanding and Applying Math
>
> Addition, as well as other mathematical concepts, although calculated formally, through an appropriate notation, are applied in a content-wise manner, regarding their meaning. Therefore, it is important to understand mathematical concepts. Without understanding, they are inapplicable. Unlike formal procedures, understanding is a human activity in which computers cannot replace us (I hope!).

1.5   Proving

Up to now we have described an important relation between numbers (their comparison) and an important operation with numbers (the operation of addition), and we found the formal rules of calculation for them. However, math does not consist only of imagining relations and operations with numbers and determining the formal rules of calculation. *It is very important to know also the properties of relations and operations: laws that are valid in the world of numbers.* For example, it is not the same whether the sum $a + b$ of any numbers a and b is equal to the sum $b + a$ or is not equal. The basic laws will maybe seem so obvious that you will ask yourself why they need to be mentioned at all. However, this is precisely why they are important – because they are the basic ones, they express some essential properties of numbers and they are the basis for discovering the more complex laws. For instance, among the basic properties of relation of comparison ($<$) are the following two properties:

1. *For any numbers a and b, either $a < b$ or $b < a$ or $a = b$.*

2. *For any numbers a, b and c, if $a < b$ and $b < c$ then $a < c$.*

The relation $<$ between numbers has been described by the comparison of sets. By using this description we could discover that the previous two properties, as well as many others, are valid. Let us, for instance, prove the first basic property. Connecting into pairs the elements of set A (with a elements) and set B (with b elements), we find that precisely one of the following situations will occur:

- All elements from set A are taken, and some elements of set B remain – then $a < b$.

- All elements from set B are taken, and some elements of set A remain – then $b < a$.

- All elements both from set A and from set B are taken – then $a = b$.

Thus, for every number a and b, either $a < b$ or $b < a$ or $a = b$.

Example 1.5.1. Let us prove the following properties of the relation $<$:

1. For any numbers a and b, if $a < b$ then $b < a$ is not the case.

2. For any number a there is a number b such that a is smaller then b, that is $a < b$.

Solution.

1. We will prove this property in two ways: (i) using the description of the relation $<$ by means of comparison of sets, (ii) using the previously mentioned basic properties.

   (i): Let $a < b$. This means that in pairing the elements of set A (with a elements) and set B (with b elements) all elements of set A are taken, and some elements of set B remain. Therefore, it cannot be that $b < a$ since it would be precisely an inverse situation — that all elements of set B are taken, and some elements of set A are left.

   (ii): Now we will provide a proof that invokes the first basic property mentioned above:

   for any numbers a and b, either $a < b$ or $b < a$ or $a = b$.

   This property says that always only one of the three mentioned options is fulfilled. Thus, if $a < b$ then it cannot be that $b < a$ (nor $a = b$).

2. If we add one more element to set A (with a elements), we will get a new set B with a bigger number of b elements. Thus, there is a number b such that $a < b$.

   Unlike the previously stated property, this property cannot be proved on the basis of the two basic properties mentioned. It can even be proved that it cannot be proved from them, but we will leave such math gems for *Circle 4*. Therefore, this last property should be added to the list of the basic properties. $\square$

In somewhat "more serious" math some properties are accepted as basic, and all others are **proved** from them. These basic properties are called **axioms** (Greek *axioma* — that which is thought worthy), and all the others – that can be proved on their basis – are called **propositions** (Lat. *proponere* — *consider*), or, if we consider them especially important, **theorems** (Greek *theorema – perceived*).

> ### The Meaning of Axioms and Proofs
>
> In the axiomatic approach, the question of the truth of all statements is reduced to the question of the truth of axioms and correctness of proofs. This gives the mathematical story greater precision, certainty, and efficiency, and reveals the logical connection of its parts.
>
> You can understand axioms as basic components of a narrative, from which we are trying to infer everything that happens in the story.

> *Euclid (fl. 300 BCE) wrote the book Elements, even today the most successful book in the history of math, in which all the known laws of geometry are derived from a small set of axioms. Such an approach has become a model for the precise and logical presentation of mathematical and other theories.*

We will leave the axiomatic approach until *Circle 4*, and here we will approach math in an "upbeat" way, "with a smile". We will emphasise some basic properties, but possibly not all. Some properties will be proved, but not all. The proofs will sometimes be rather more like persuasions than actual proofs. We are not going to be particularly precise in our proofs now because the main subject of our studies is numbers, and everything else is an auxiliary tool in the study. At the moment, we are interested in the counting of the CDs, and not the question of what the correct proof is. Of course, by reading the proofs made by others, and making one's own, one acquires the experience of proving and sharpens one's thinking. This is also a very important element of math education. As people in my region would say in jest: in the end, you will have such sharp thinking, you will be able to use it to cut watermelons.

However, it's important to know that to someone who claims something, at least in math, you can always say *prove it!*, whether this person is, your math professor or *Her Majesty* (OK, be careful with the *Great Leader*). This is equality in math. But when you claim something yourself, be prepared to try to prove it using the basic ideas of the math story. This is responsibility in math. Of course, math is not just responsibility but also an adventure. Therefore, let your imagination go and set any **hypotheses** (Greek *hipotithenai – assume*: statements whose truth – or non-truth – has not been proved) about numbers, whether you know how to prove them or not. Maybe you will solve some of these hypotheses by yourself or with help of others; maybe with the internet and books. Some you will not be able to solve even in this way. No wonder! There are even today some simple hypotheses about numbers, set several hundred years ago, that have not been solved yet (there are even "worse" things, but this will also be discussed in *Circle 4*).

1.6   Basic Properties of Comparison

> **Basic Properties of Comparison of Natural Numbers**
>
> 1. *For any number a there is the first number s such that a is smaller than s* (it is called the **immediate successor** of number *a*), *and unless a = 0 then there exists also the first number p, which is smaller than a* (and it is called the **immediate predecessor** of number *a*) (**the existence of the immediate successor and predecessor**)
>
> 2. *Zero is the smallest number, i.e. for every number a, unless a = 0 than it is true that* $0 < a$ (**the existence of the first number**)
>
> 3. *For any numbers a, b and c, if* $a < b$ *and* $b < c$ *then* $a < c$ (**transitivity**)
>
> 4. *For any numbers a and b, either* $a < b$ *or* $b < a$ *or* $a = b$ (**strict comparability**)
>
> 5. *In every non-empty set of numbers there is a number which is the smallest among them* (**the existence of the smallest number in a set of numbers**)

In this way, we have precisely defined the basic idea about natural numbers: we imagine them as a sequence of items in which there is the first member, and after every member there is the new next member:

$$0, 1, 2, 3, 4, 5, \ldots$$

For example, the immediate successor of the number 4 is the number 5, and the immediate predecessor of the number 4 is the number 3. Numbers smaller than 4 are to the left of 4 (they appear earlier in the series) and the numbers which 4 is smaller than are to the right of 4 (they occur later in the series). For such numbers we also say that they are **bigger** than 4. However, *to be bigger* ($>$) is a different relation than *to be smaller* ($<$). Of course, they are closely connected because $a > b$ precisely when $b < a$. Simply, we read the relation from right to left instead of from left to right. Here the relation $>$ is naturally imposed on us. Its meaning can be described by the relation $<$. The procedure of introducing new symbols whose meaning is described by symbols already introduced is called *defining*, and the next topic is dedicated entirely to it.

1.7   Defining

Sometimes another kind of comparison of natural numbers is introduced. It is not so content-full as the comparison that we have already introduced, but it has proved to be easier to work with. In math terminology it is said that it is *technically more comfortable*.

Less than or Equal Relation

We say that $a \leq b$ (**a less or equal b**) precisely when $a < b$ or $a = b$.

Thus, for example $3 \leq 4$ since $3 < 4$, but also $3 \leq 3$ because $3 = 3$.

Let's note that in mathematical language the phrase *if and only if* (abbreviation: *iff*) is commonly used instead of *precisely when*.

Such description of a new symbol, word or concept using already introduced symbols, words and concepts is called **definition** (lat. *definere – set bounds to*). The definition precisely determines the meaning of the new symbol, word or concept using the meaning of old symbols, words, and concepts. Thus, in the previous definition it has been precisely determined what it means that $a \leq b$ according to what it means that $a < b$ and $a = b$. On the basis of the definition we can check whether some numbers are in the defined relation. In this way we already determined that $3 \leq 4$. Also, based on the definition we can determine the properties of the relation $\leq$ as well.

Properties of Less than or Equal Relation

1. *For any a, $a \leq a$* (**reflexivity**)

2. *For any a and b, if $a \leq b$ and $b \leq a$ then $a = b$* (**antisymmetry**)

3. *For any a, b and c, if $a \leq b$ and $b \leq c$ then also $a \leq c$* (**transitivity**)

4. *For any a and b, $a \leq b$ or $b \leq a$* (**non-strict comparability**)

During your schooling, I believe, you got so used to the properties of the relation $\leq$, that you find them perfectly acceptable and there is no need to convince you through proofs of their correctness. I will, however, prove the property of antisymmetry, as an illustration of the process of proof based on definitions.

Example 1.7.1. Let us prove:

For any a and b, if $a \le b$ and $b \le a$, then $a = b$.

Solution. Let a and b be such that $a \le b$ and $b \le a$. We need to prove that then $a = b$. We will prove this using the definition of the relation $\le$ and the properties of the relation $<$. According to the definition of $\le$:

1. $a \le b$ means that $a < b$ or $a = b$ 2. $b \le a$ means that $b < a$ or $b = a$

Thus, we have these four options:

1. $a < b$ and $b < a$ 2. $a < b$ and $b = a$ 3. $a = b$ and $b < a$ 4. $a = b$

However, the property of strict comparability of relation $<$ (4th basic property) says that precisely one of $a < b$, $b < a$ and $a = b$ is valid. Therefore, we reject the first three options, leaving the fourth one remaining: that $a = b$: precisely what we needed to prove. $\square$

There was nothing especially mathematical in the proof, except maybe the use of symbols (we prefer writing $\le$, than *less or equal*). It is in the same way that in crime movies your favourite inspector, using the data collected, proves the identity of the killer. Or, you prove to your friend the minimal score difference your club must win by, in order to qualify for the next round of the competition.

A more serious math storytelling requires not only an emphasis on the fundamental properties which are the basis for proving all the others, but also an emphasis on the given symbols and words that are used to compose a story, and whose meaning we understand. If we want to use new symbols and words, we must define them by the old ones. We do this by means of a definition, a description by which we specify the meaning of a new sign (word) by means of the meaning of old signs (words). We will make these descriptions, at this level, intuitively. In this way, we have already defined comparison and addition by means of concepts related to sets. By using and making definitions, you will also develop the feeling for correct definitions. We will leave the completely accurate analysis of the definition process until *Circle 4*.

> ### The Meaning of Basic Words, and Definitions
>
> In composing a mathematical story we use various words. Some of them we consider to be basic words. Usually, we understand their meaning well. By defining, the meaning of every other word is reduced to the meaning of the basic words and correctness of definitions. This makes the math story clearer – everything said has precise meaning and no ambiguity. Moreover, when we use definitions to substitute entire descriptions with one word, we simplify expression and structure thinking. Definitions are one of the basic mechanisms of abstraction that allow us to successfully handle the complexity of content.

Thus, at least in math, when somebody uses a certain word, you always have the right to ask for its definition (equality!), but also when you use a certain term, think about whether you can define it by means of the basic words (responsibility!). Of course, this is how it is in a completely *serious* approach. To us this will only represent an ideal that we will *moderately* adhere to.

1.8 Basic Properties of Addition

Now we will enumerate some basic properties of addition of numbers.

> ### Basic Properties of Addition of Natural Numbers
>
> 1. *For any a and b, $a + b = b + a$* (**commutativity of addition**)
>
> 2. *For any a, b and c, $(a + b) + c = a + (b + c)$* (**associativity of addition**)
>
> 3. *For any a, $a + 0 = a$* (**neutrality of zero**)
>
> 4. *For any a, b and c, if $a + b = a + c$, then $b = c$* (**cancellation law**)

We have described the addition $a + b$ of numbers a and b as a total number of elements of two sets A and B, of which one has a, and the other b elements, and they have no common elements. From this description the mentioned properties follow easily. Let us prove, for instance, the commutative property. In whichever way we join the elements of sets A and B, either by adding elements of set B to elements of set A, or by adding elements of set A to elements of set B, eventually we will have together the same elements, and their number will also be the same. Thus, $a + b = b + a$. Through similar thinking we could prove other properties as well.

Because of the property of associativity we need not write the brackets that determine the order. Instead of writing $1 + (2 + 3)$ or $(1 + 2) + 3$, we write $1 + 2 + 3$. Whichever way we add, we will always get 6. Due to commutativity we may even change the order of addition. We can, for instance, add first 1 and 3, and only then add 2. We will obtain 6 again. Of course, it is easier to use the simpler order.

Example 1.8.1. Let us calculate $(13 + 29) + (37 + 21)$.

Solution. In the expression $(13 + 29) + (37 + 21)$ the brackets determine the order of calculation with which we will identify the described number. The first and the third addition can be performed because we know which numbers are added, but the second addition cannot be done because we do not know yet which numbers are added. Since $13 + 29 = 42$ and $37 + 21 = 58$, this yields

$$(13 + 29) + (37 + 21) = 42 + 58 = 100$$

However, due to the properties of commutativity and associativity we can forget about brackets and calculate in any (possibly simpler) order:

$$(13 + 29) + (37 + 21) = 13 + 29 + 37 + 21 = (13 + 37) + (29 + 21) = 50 + 50 = 100$$

$\square$

The connection between comparison and addition is as follows:

Compatibility of Comparison and Addition

For any a, b and c, $a + c < b + c$ precisely when $a < b$
(We say that addition of a number **preserves** the comparison.)

1.9 Subtraction

Let us imagine that Uninterested took his 19 CDs from the joint collection of 65 CDs and took them with him for a summer vacation. How many CDs are left? The taking away of elements from a set is reflected as the operation of subtraction of numbers associated with these sets.

Subtraction of Natural Numbers

Let there be a elements in set A and let b elements be taken from it (there must be $b \leq a$). The number of remaining elements is called **difference of the numbers** a and b and we write $a - b$ (read ***a minus b***). Symbol '$-$' (*"minus"*) is the sign of the operation that is called **subtraction**.

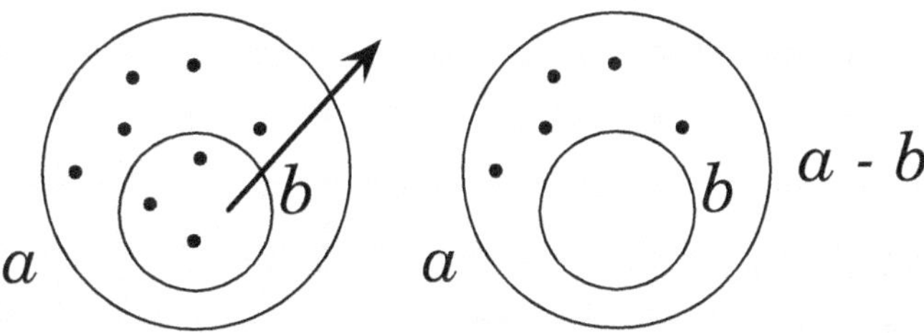

In the sequence of numbers we will obtain the number $a-b$ by starting from number a for b units in the direction of smaller numbers:

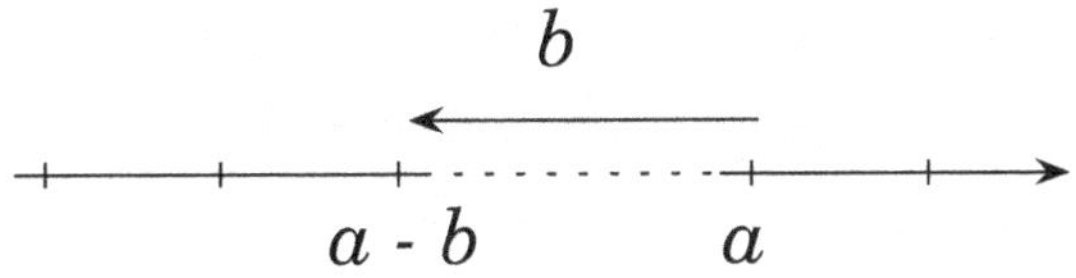

Thus, $65 - 19$ CDs remain. What number is that? The decimal notation gives again a simple formal rule that we have known since school:

$$\begin{array}{r} 65 \\ -19 \\ \hline 46 \end{array}$$

We will first subtract the ones. However, since we cannot subtract 9 from 5, we will borrow one of the 6 tens. Thereby, we will get 15 ones. After we subtract 9 from them, we get 6 ones. Due to the borrowing procedure we have 5 tens left. We subtract 1 ten from them and obtain 4 tens. Thus, the result is 46. It is clear that Astonished has now an equal number of CDs as before the two friends' CDs were combined. This refers to one more obvious property: that adding and subtracting are mutually opposite or inverse operations. If we add a certain quantity, and then subtract it, we will obtain once more the original quantity.

> **Inverse Relationship between Addition and Subtraction**
>
> 1. *For any a and b, $(a + b) - b = a$*
>
> 2. *For any a and b, if $a \geq b$ then $(a - b) + b = a$*
>
> 3. *For any a, b and c such that $a \geq b$,*
>   *$a - b = c$ precisely when $c + b = a$*

If we know that when the CDs were combined there were $46 + 19 = 65$ CDs, we can conclude even without calculation that Astonished will have 46 CDs when Uninterested takes away his 19 CDs.

Let us note that we have defined the subtraction $a-b$ of natural numbers a and b only if $a \geq b$. Thus, it has not been defined what expression '$3-5$' means (it has no meaning, for the moment).

1.10 Multiplication and Division

In the *Happy Childhood* primary school, a window in a classroom has been broken. Not knowing the identity of the culprit, the headmaster requested that every schoolchild from the class pay 45 kuna for the damage (kuna is the Croatian currency, the word *"kuna"* means *"marten"* in Croatian, and commemorates the use of marten pelts as units of value in medieval trading). The mathematically-aware schoolchildren have decided to first check whether his request is in order. The class consists of 32 schoolchildren and every one has to pay 45 kuna. This is a total of $45 + 45 + 45 + \ldots$ No, no. The schoolchildren have just learned about multiplication and now know that they can calculate this in a faster way. There are 32 groups of 45 kuna each.

Multiplication of Natural Numbers

If we have b sets of a elements each, with no joint elements, the total number of elements is called **product of the numbers** a and b and we write $\boldsymbol{a \cdot b}$ (read $\boldsymbol{a}$ **times** $\boldsymbol{b}$ or $\boldsymbol{a}$ **multiplied by** $\boldsymbol{b}$). The symbol '·' (*"times"*) is the mark of the operation that is called **multiplication**.

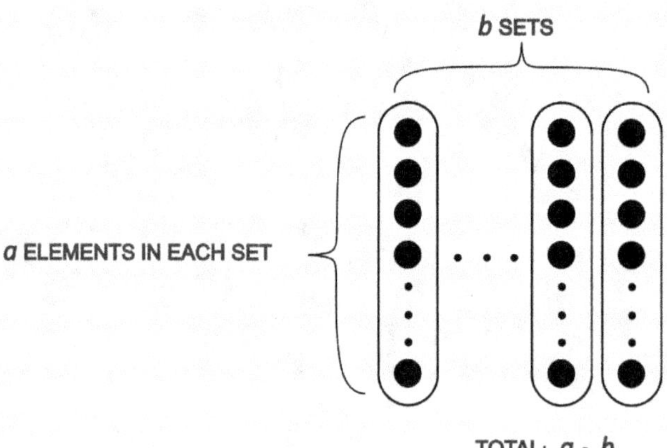

Multiplication can be described by means of addition as well:

$$a \cdot b = \underbrace{a + \ldots + a}_{b \text{ times}}$$

Here again, as in the case of addition, one must keep in mind that the sets *shall not have common elements*.

Since primary school we have known about the formal procedure of how to use the decimal notation of numbers to calculate their product. Now, we will use this procedure to calculate how much $45 \cdot 32$ is:

$$
\begin{array}{r}
45 \cdot 32 \\
\hline
135 \\
+90 \\
\hline
1440
\end{array}
$$

This is the formal procedure for multiplication as learned in Croatia. Maybe you have learned a different procedure for multiplication. Whichever procedure, it is important that it is correct, that is to say, that the result is precisely the product of the numbers. Analyse your multiplication procedure and try to determine why it is correct. Meanwhile, I will just pop over to the glazier Tonko to ask about the cost of the installation of the new glass for the window. Puff, puff, pant. Here I am back again. The price will be 500 kuna at the most. This is much less than the headmaster's request. How much should each schoolchild pay for the broken glass? The problem is reversed now. How to divide 500 kuna into 32 groups so that there is an equal number of kuna in each of them? Sometimes it is good strategy in solving problems to try to solve a similar, but smaller, problem – how to divide 500 kuna into 32 groups so that each one consists of an equal number of kuna. No, I did not have this kind of reduction in mind, but rather a reduction of the size of the numbers – for example, how to divide 7 kuna into 3 groups so that there is an equal number in each.

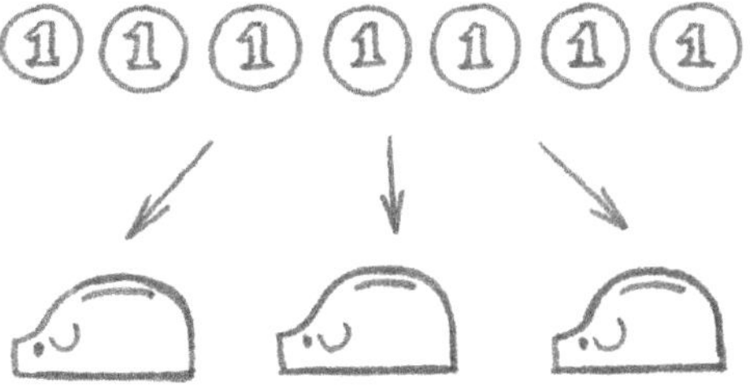

If we put into each group 1 kuna, we have distributed 3 kuna, and 4 remain unallocated:

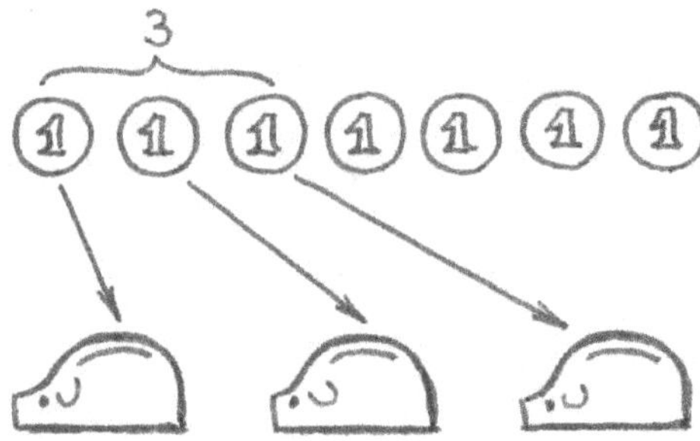

Let us allocate another 3 kuna in the same manner. Now we have divided $2 \cdot 3 = 6$ kuna. There are 2 kuna each in each group, and one still remains unallocated.

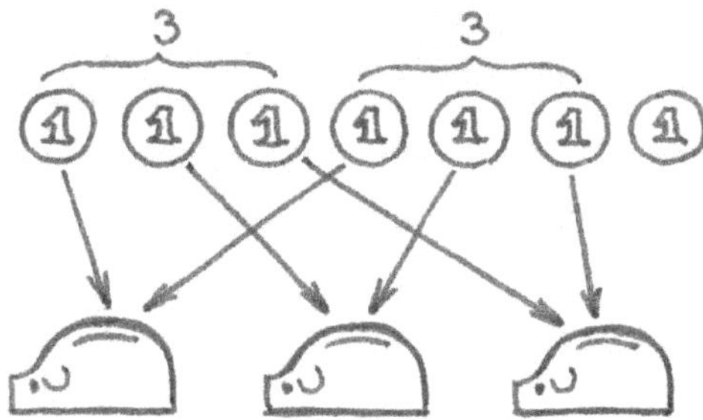

However, we cannot repeat the procedure, since only one is left. We can only continue dividing it into three groups if we convert it into lipas (a kuna is made up of 100 lipa: the word *"lipa"* means *"linden (lime) tree"*). Such a division of items into parts will be measured later by rational numbers. However, in the world of natural numbers we distribute entire items, not their parts. Therefore, the procedure stops here. Seven kuna is divided equally into 3 groups so that there are 2 kuna in every group. Two is the maximal number of kuna that multiplied by 3 does not yield more than 7 ($2 \cdot 3 \leq 7$). Here, one more kuna remains unallocated. This is the remaining (smaller than 3) number of kuna that needs to be added to the above-mentioned product so that the total number of kuna is obtained: $2 \cdot 3 + 1 = 7$.

We could have applied this procedure to any number of kuna and groups. The conclusion would be the same.

Division with Remainder of Natural Numbers

Euclid's theorem. *For any two natural numbers a and b (b cannot be 0) there are unique natural numbers n and m such that*

$$a = n \cdot b + m \text{ and } m < b$$

The number n is called the **result or quotient of division with remainder** of numbers a and b and we mark it $a \div b$. The number m is called the **remainder** and we mark it $a \bmod b$ (lat. *modulo – up to measure*). Division with remainder is also called **integer division**.

The quotient and the remainder can be found by a trial and error method:

Finding the Quotient and Remainder of Integer Division

In order to find the result and remainder of the division of number a by number b we guess the biggest number that multiplied by b still yields a number that is not greater than a. This is precisely $a \div b$.

It is easy to find the remainder: $a \bmod b = a - (a \div b) \cdot b$.

This procedure applied to the division of the number 7 by the number 3 yields the following. Since 2 is the biggest number that multiplied by 3 yields no more than 7, then

$$7 \div 3 = 2$$

We calculate:

$$7 \bmod 3 = 7 - (7 \div 3) \cdot 3 = 7 - 2 \cdot 3 = 7 - 6 = 1$$

However, let us return to the headmaster. We have to divide 500 kuna (the cost of replacing the broken window) equally between 32 schoolchildren in the class. This is precisely $500 \div 32$: the biggest number that multiplied by 32 yields a number smaller or equal to 500. By guessing, after several trials we will find out that this is 15. Indeed, $15 \cdot 32 = 480$, and up to 500 only 20 kuna remain ($= 500 \bmod 32$). We can calculate it more simply by the procedure of division with remainder that you know from your primary school, and which in this case looks as follows :

$$
\begin{array}{l}
500 : 32 = 15 \\
\underline{-32\downarrow} \qquad\quad \parallel \\
\quad 180 \qquad\; 500 \div 32 \\
\underline{-160} \\
\quad\; 20 = 500 \bmod 32
\end{array}
$$

I leave it to you to think about why this procedure is correct (or if you learned another one, why this other one is correct). But now we have to do something more important. We have to find out who will pay the remaining 20 kuna that cannot be equally divided any more (without conversion into lipas). Bingo! – I've got it! We will ask the glazier Tonko to reduce the price for the work, since we are advertising him in this story. Thus, every schoolchild will pay the headmaster 15 instead of 45 kuna, as he had requested before. And who can say now that math is not useful!

Example 1.10.1. Let us calculate

1. $15 \div 6$, $15 \bmod 6$ 2. $6 \div 15$, $6 \bmod 15$ 3. $6 \div 6$ 4. $6 \div 0$

Solution.

1. Since $0 \cdot 6 = 0$, $1 \cdot 6 = 6$, $2 \cdot 6 = 12$, $3 \cdot 6 = 18$, $\ldots$, then $15 \div 6 = 2$.

   Since $2 \cdot 6 = 12$, then $15 \bmod 6 = 15 - 12 = 3$.

2. Since $15 > 6$, then $6 \div 15 = 0$, and $6 \bmod 15 = 6$.

   This example can be easily generalised: whenever the divider is greater than the divisor ($b > a$), the following is valid: $a \div b = 0$ and $a \bmod b = a$.

3. Since $6 = 1 \cdot 6$, then $6 \div 6 = 1$. Generally, since $a = 1 \cdot a$, then $a \div a = 1$.

4. $6 \div 0$ should be the biggest number that multiplied with 0 yields not more than 6:

   $1 \cdot 0 = 0$, $2 \cdot 0 = 0$, $3 \cdot 0 = 0, \ldots$

   It would be a pity to spend the rest of our lives continuing this series. It is better to notice that every number multiplied by zero yields zero, thus less than 6, so there is no biggest such number. Thus, $6 \div 0$ does not exist, just as $a \div 0$, or $a \bmod 0$ do not exist either. For any a there is no dividing by zero. This is nothing uncommon. Some formally correct descriptions simply make no sense. Instead of naming just one item they do not name anything. When I say 'your nose', this is the description of your nose, but when I say 'the nose of our town', this means nothing. Thus, expression '$6 \div 2$' is the description of a number, the number 3, but '$6 \div 0$' does not mean anything. For such descriptions we say that **they have no meaning**, in other words, that what they describe **does not exist**. In diplomatic mathematical language one may say that **they are not defined** (it seems that the mathematical community do not like to mention senselessness too much). The operation of division is conceived in such a way that division is impossible when the divisor is zero. Therefore (and for no other reason), there is no dividing by zero. $\quad\square$

Let us note once again that decimal notation has allowed us such simple algorithms for calculating basic operations with natural numbers. Before decimal notation, Roman notation was used in Europe. It was much more difficult to calculate in that notation, especially when multiplying and dividing numbers. The changeover to decimal notation in the 12th to 15th century period meant a huge advance in math. *This example illustrates the general rule, which is worth mentioning once again, that mathematical notation is a very important part of math.*

> *Fibonacci (1170–1240) popularised the Hindu-Arabic numeral system in the Western world primarily through his composition in 1202 of Liber Abaci (Book of Calculation). The book was well-received throughout educated Europe and had a profound impact on European thought. Today, the book is famous because of an example that Fibonacci introduced: the so-called sequence of Fibonacci numbers, which appears very interesting.*

You can read more about the meaning of basic arithmetic operations and calculation procedures at `https://en.wikipedia.org/wiki/Elementary_arithmetic`.

1.11 Multiplication

And while the headmaster is leaving this story happy that the schoolchildren have managed division with remainder well, let us look at how multiplication can sometimes help us in counting.

Example 1.11.1. If you have 5 T-shirts and 3 pairs of trousers, in how many different combinations can you dress yourself? And what if you have 5 T-shirts, 3 pairs of trousers and 2 pairs of shoes?

Solution. With each of the 5 T-shirts we can put on one out of the 3 pairs of trousers, and the total number of outfits is $5 \cdot 3 = 15$. If we also have two pairs of shoes, for every choice of T-shirt and trousers we have two more outfits, depending on which shoes we select. Since there are 15 ways of combining T-shirts and trousers, the total number of outfits is $15 \cdot 2 = 30$. $\square$

This example can easily be generalised.

> ### Principle of Counting by Multiplication
>
> If we can select the first item in a ways, and in every choice of the first item, the second item can be chosen in b ways, then the total number of selections is $a \cdot b$.

Example 1.11.2. The teacher sent Johnny, George and Stevie to fetch him the register, a sponge and a chalk. To avoid getting tired, each one of them will carry one item. How many different carrying options do they have?

Solution. Let's say Johnny is the first to choose. He has 3 options. He can take either the register, or the sponge or the chalk. Whichever Johnny chooses, George can choose one of the two remaining items. Whatever Johnny and George select, Stevie will have no options because he will have to take the only remaining item. Thus, according to the principle of counting by multiplication, the total number of options is $(3 \cdot 2) \cdot 1 = 6$. $\square$

The basic properties of multiplication are as follows:

Basic Properties of Multiplication of Natural Numbers

For any natural numbers a, b and c it holds that:

1. $a \cdot b = b \cdot a$ (**commutativity**)

2. $(a \cdot b) \cdot c = a \cdot (b \cdot c)$ (**associativity**)

3. $1 \cdot a = a$ (**neutrality of one**)

4. $0 \cdot a = 0$

5. *For $a \neq 0$, if $b \cdot a = c \cdot a$ then also $b = c$* (**cancellation law**)

6. $a \cdot (b + c) = a \cdot b + a \cdot c$ (**distributivity of multiplication over addition**)

7. *For $a \neq 0$, $b \cdot a < c \cdot a$ precisely when $b < c$* (**compatibility of multiplication and comparison**)

You needn't remember the names of these properties. Your math aptitude will not depend on whether you know their names or not. But if you wish to participate in *Who Wants to Be a Millionaire* or some similar game show, then the situation changes. The only important thing is that you know these names – everything else is irrelevant. And one more thing. When it is clear what is meant then we will prefer to write ab instead of $a \cdot b$.

Some of the above-mentioned properties are obvious, e.g. the 3rd and 4th properties, and these are easy to prove from the definition of multiplication. Some, however, are not so obvious, for example the 1st and 5th. If, for instance, we take 3 times 4 kuna, why should this be the same as if taking 4 times 3 kuna? All the properties we have looked at so far have been rather obvious. I have proved some for the sake of illustrating the proof procedure in math, rather than convincing you of their correctness. However, *the real*

power of proof is seen precisely in situations when we are not sure whether a certain claim about numbers is true or not. Maybe this is just one such situation. It is easy to ascertain from the multiplication table or using a calculator that $4 \cdot 3$ is the same as $3 \cdot 4$. However, this proof does not verify that this is always the case, for any two numbers. Therefore, I will present a proof that can be easily applied to any two numbers. According to the definition of multiplication, $4 \cdot 3$ is the total number of elements from 3 sets with 4 elements each, provided they have no common elements:

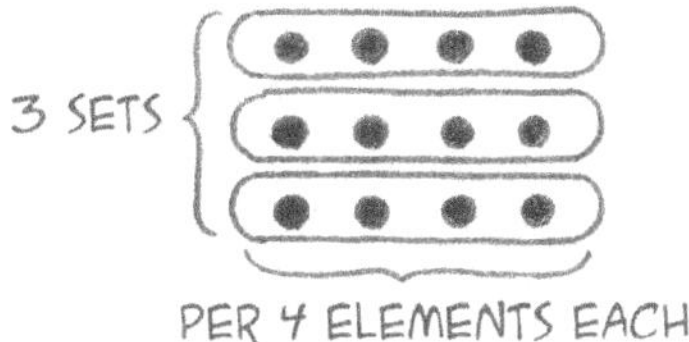

If we take the first elements (first column) from each set, we will get a set of 3 elements. In the same way the other elements of the sets (second column) will yield the second such set, etc.:

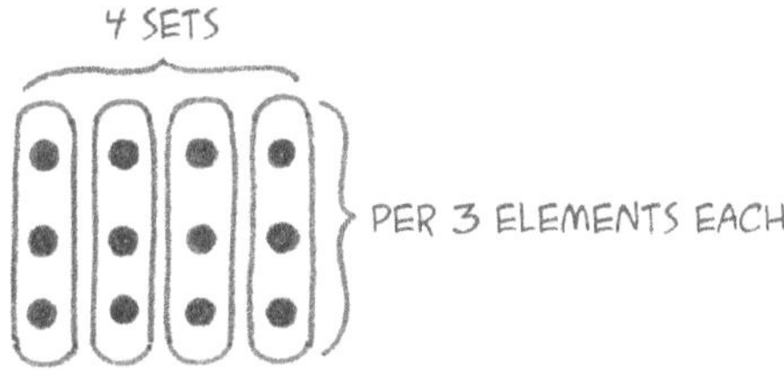

Thus, from the initial 3 sets with 4 elements each, by redistribution we formed 4 sets with 3 elements each. This means, according to the definition of multiplication, that $4 \cdot 3 = 3 \cdot 4$. *In this example we see that we can prove the same claim in many ways. They are all equally valid in effect – they prove that something is so. But some proofs teach us even more. They teach us to understand why something is true and show us what else there may be.* This is precisely such a proof. It is easy to apply it to any two numbers. If we replace the concrete numbers 3 and 4 with any numbers a and b, we obtain the proof of the general property of commutativity of numbers. This proof shows us the manner of the redistribution of elements from b of mutually separated sets with a elements each in a separated sets with b elements each. Thus, it shows us that $a \cdot b = b \cdot a$.

In the same way the following figure shows that the distributivity law is valid:

$$a \cdot (b + c) = a \cdot b + a \cdot c$$

In successive multiplication, as in addition, the order of multiplication is not important. If we combine multiplication and addition then we have to use parentheses to determine the order of operations.

Example 1.11.3. Let us evaluate: 1. $(2+3) \cdot 4$,   2. $2+(3 \cdot 4)$,   3. $2+3 \cdot 4$

Solution.

1. The description of the number determines the order of operations:

    $(2+3) \cdot 4 = 5 \cdot 4 = 20$

2. $2+(3 \cdot 4) = 2 + 12 = 14$

3. In this description there are no brackets that determine the order of operations. The previous two examples show that different orders of calculation yield different results. Therefore, this description is ambiguous. We can compare this with the example of when somebody asks me who my best friend in the neighbourhood is. I say *Štef* (the Croatian variant for Stevie), and in my neighbourhood there are seven people with that name Štef (it is a common name in the part of Croatia where I live). To make the description precise, so as to use it to identify only one person, I have to add his family name and maybe even his father's name. Just as in the previous description we have to add brackets or introduce an agreement on what is meant when there are no brackets. And **the agreement on the order of operations** is, as you know, *in an ambiguous situation, to do first multiplication and division (arithmetic operations of the second order), and then addition and subtraction (arithmetic operations of the first order)*:

    $$2+3 \cdot 4 = 2+(3 \cdot 4) = 2 + 12 = 14$$

    You got used to this in primary school, and I have already applied this agreement, for example, when I stated the law of distributivity.   □

1.12 Calculating and Simplifying

Example 1.12.1.

1. How much is $345 \cdot 3 + 345 \cdot 7$?

2. How much is $346 \cdot 3 + 345 \cdot 7$?

3. Let us analyse the procedure of the addition 46 + 19 using the meaning of decimal notation and the laws of arithmetic operations.

Before we solve this example, let's dwell a little on the meaning of these questions. For example, what does the question "How much is $345 \cdot 3 + 345 \cdot 7$?" mean at all? The linguistic form (expression) '$345 \cdot 3 + 345 \cdot 7$' is the description of a number. *Every number has its standard name in the decimal notation and our aim is to obtain from its description its standard name (standard description).* The same thing happens in everyday life. When you do not know a female person, but you know her son John, then you will identify this person by describing her as 'John's mother'. Of course, this person has a standard name and the aim of the description is to find out from that description the person's standard name. *When a description is a little more complex, then the procedure of finding the standard name is called calculation.* Calculation is based on the following simple observation: when we place the sign of equality between two descriptions, this means that they describe the same object. So $2 + 4 = 2 \cdot 3$, as *father of Johnny = author of this book*. In a more complex description, whenever we find a part that is also a description of an object, we can replace it with another description of that object, possibly with its standard name. *Calculation of a description consists precisely of a series of such replacements until we obtain the standard name.*

What It Means to Calculate a Description

Every number has its standard name in the decimal notation and our aim is to obtain from its description its standard name.

Calculation of a description is a sequence of transformations of the description such that every transformed description describes the same number, until we obtain the standard name of the number.

Solution.

1. By calculating the parts of the description in order, we will calculate the whole description:

$$345 \cdot 3 + 345 \cdot 7 = 1035 + 2415 = 3450$$

However, according to the distributivity law, this description can be replaced by another description of the same number which is easier to calculate:

$$345 \cdot 3 + 345 \cdot 7 = 345 \cdot (3 + 7) = 345 \cdot 10 = 3450$$

.

> **The Procedure of Finding a Number from its Description**
>
> The procedure of finding a number from its description consists of a series of simple steps. In each step a part of the description is identified which is in itself a description of a number. It is replaced by a different description of this number. We can calculate directly – we replace a part of the description that consists of an operation with known arguments (e.g. $345 \cdot 3$) with the result (1035). Or we can transform a part of the description (in the previous case the entire description $345 \cdot 3 + 345 \cdot 7$) into a new description ($345 \cdot (3 + 7)$) of the same number, applying some property of operations (in this case the law of distributivity).

2. Instead of direct calculation, we will again transform a little to obtain the result more easily:

$346 \cdot 3 + 345 \cdot 7 =$ (we will disassemble the first number)

$(1 + 345) \cdot 3 + 345 \cdot 7 =$ (we will apply distributivity)

$1 \cdot 3 + 345 \cdot 3 + 345 \cdot 7 =$ (we will apply distributivity)

$1 \cdot 3 + 345 \cdot (3 + 7) =$ (direct calculation)

$3 + 3450 = 3453$

3. $46 + 19 =$ (in terms of decimal notation)

$4 \cdot 10 + 6 + 1 \cdot 10 + 9 =$ (we will add the ones)

$4 \cdot 10 + 1 \cdot 10 + 15 =$ (we will isolate the tens)

$4 \cdot 10 + 1 \cdot 10 + 1 \cdot 10 + 5 =$ (we will add the tens)

$(4 + 1 + 1) \cdot 10 + 5 =$ (in terms of decimal notation)

$6 \cdot 10 + 5 = 65$ ☐

We have already seen how decimal notation allowed simple algorithms for arithmetic operations with numbers. In previous calculations, we found another characteristic of the mathematical language to be of great help – **symbolisation**.

> ### The Importance of Symbolisation in Mathematics
>
> The introduction of symbols instead of words for naming numbers and operations and relations between numbers made mathematical descriptions not only much more summarised regarding descriptions in natural language, but it also gave the calculation procedure a simple formal character.

You can best assure yourself of this by translating a complex mathematical description into natural language and by trying to calculate it in the natural language.

Example 1.12.2. Two coaches of a football club, *Our Club FC*, took a football team of 27 little guys on a seven-day summer trip to the seaside. With a youth hotel called *Our Kids* they agreed on a full board price of 100 kuna per day per child. Rent of the football field cost 250 kuna per day, and a return bus ticket was 200 kuna per child. The parents were given a price of 1432 kuna per child, justified by the quality of the hotel and field, and the claim that the future football development of their children would be almost impossible without these preparations. I must say here that the coaches were indeed honest in one regard. They were fair in sharing their realised *profit*: each of them got half. How much money did each of the coaches earn?

Solution. The total costs were as follows. Every child spent $7 \cdot 100 = 700$ kuna for the accommodation and an additional 200 kuna for the trip. This is a total of 900 kuna. As there were 27 children, the total amount is obtained by multiplying $900 \cdot 27 = 24\,300$ kuna. We should add here the seven-day rent for the field, which was $7 \cdot 250 = 1750$ kuna. Thus, the total outlay was $24\,300 + 1750 = 26\,050$ kuna. The parents gave the coaches a total of $1432 \cdot 27 = 38\,664$ kuna. Thus, they trousered $38\,664 - 26\,050 = 12\,614$ kuna. After splitting the proceeds each of the coaches got $12\,614 : 2 = 6307$ kuna. $\square$

Example 1.12.3.

1. Let us simplify $2a + 3b + 4a + 5b$.

2. Let us simplify $2a \cdot (3 + 2b) + 3b \cdot (3a + 2)$.

3. Let us prove that $(a + b) \cdot (c + d) = ac + ad + bc + bd$

4. Let us simplify $(2a + 3) \cdot (2 + 5b + 3c)$.

Solution.

1. Before we solve this example we will again reflect on the purpose of
what is required. What does it mean to simplify e.g. $2a + 3b + 4a + 5b$?
What is $2a + 3b + 4a + 5b$ anyway? What the number $2a + 3b + 4a + 5b$ is,
depends on the numbers a and b. Only when we have specified what a
and b are, can we specify the number $2a + 3b + 4a + 5b$, by calculation.
For instance, if $a = 2$ and $b = 3$, this is the number

$$2a + 3b + 4a + 5b = 2 \cdot 2 + 3 \cdot 3 + 4 \cdot 2 + 5 \cdot 3 = 4 + 9 + 8 + 15 = 36$$

However, as long as a and b are not specified, until these are a *certain* a and b, the number $2a + 3b + 4a + 5b$ remains undetermined –
it is a *certain* $2a + 3b + 4a + 5b$. Such a description is called **open description** (descriptions from the example 1.12.1 are **closed descriptions**). From this description we cannot identify what number it describes since it depends on the numbers a and b. We can only find a
simpler description of the number (or a different description which we
find better for some reason). Here, direct calculation can very rarely
help us. For instance, we cannot calculate how much $2a$ is, when it is
not specified how much a is (let us not forget that $2a$ is a shorter notation for $2 \cdot a$). Instead, we have to apply the properties of operations in
order to replace a part of the description with a different description of
the same number:

$2a + 3b + 4a + 5b =$ (we will apply commutativity of addition)

$2a + 4a + 3b + 5b =$ (we will apply distributivity)

$(2 + 4)a + (3 + 5)b =$ (we will calculate directly)

$6a + 8b$

The description cannot be simplified any further.

This example illustrates the general situation.

> **The Procedure of Simplification of an Open Description**
>
> We simplify open descriptions (or transform them into a description which we find better for some reason) in a series of simple steps. In every step we look at the description and observe the part which is itself a description of a number (e.g. $2a + 4a$ from the previous example). We replace it with a different description of the same number ($2a + 4a \mapsto (2 + 4)a$), regardless of what number this is. Usually, we transform this part according to a certain property of the operation (in this case according to distributivity), and sometimes we can calculate it directly ($(2 + 4)a \mapsto 6a$). We usually transform the descriptions in the same order in which we would calculate directly if we knew all the elements of the description.

2. In the order in which we would calculate directly if we knew a and b, we simplify the description:

   $2a \cdot (3 + 2b) + 3b \cdot (3a + 2) =$ (distributivity)

   $2a \cdot 3 + 2a \cdot 2b + 3b \cdot 3a + 3b \cdot 2 =$ (commutativity of multiplication)

   $6a + 4ab + 9ab + 6b =$ (distributivity)

   $6a + 13ab + 6b$

3. For example, when in the first task we found that $2a + 3b + 4a + 5b = 6a + 9b$, we obtained more than just an ordinary equality of two numbers. Since in the calculation we did not use anything specific about a and b, except their being certain numbers, we determined that this is valid for *for any number a and b. So we obtained a *general statement* about numbers, a statement that is valid for all numbers. This is far more than a mere equality, e.g. that $2 + 2 = 1 + 3$, which is only a *specific statement* about the number 4. *It is very important to know how to think about the arbitrary numbers $a, b, c \ldots$. This manner of thinking reveals general laws of the world of numbers.* By simplifying the description $(a + b) \cdot (c + d)$ we will determine one such law:

$(a+b)\cdot(c+d) = \quad$ (we will apply distributivity)

$(a+b)\cdot c +(a+b)\cdot d = \quad$ (we will apply distributivity)

$a\cdot c + b\cdot c + a\cdot d + b\cdot d =$

(we will apply commutativity of multiplication and addition)

$ac + ad + bc + bd$

Formally, the total effect of "multiplying brackets" is multiplication of every member from one bracket by every member from the other bracket. In the same way one may prove that the same is valid in various other combinations, for example, when within the brackets are several members that we add or subtract.

"Multiplying Brackets" Procedure

The formal rule of "multiplying brackets" in which there is addition of numbers says: every member of the addition from one bracket is multiplied by every member of the addition from the other brackets and the obtained products are summed:

$$(a+b)\cdot(c+d) = a\cdot c + a\cdot d + ab\cdot d + b\cdot d$$

---

4. According to the rule of multiplying the brackets, we obtain:

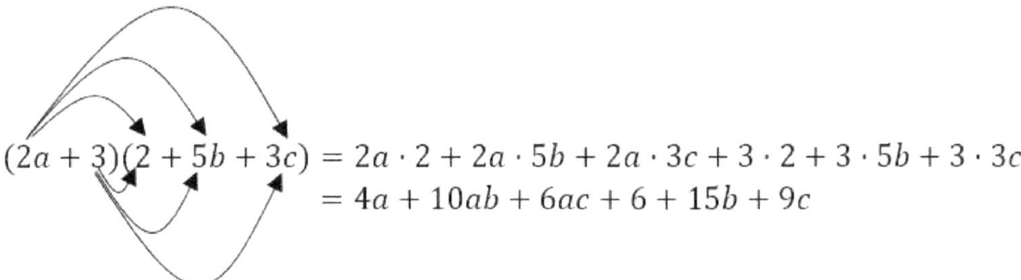

$$(2a+3)(2+5b+3c) = 2a\cdot 2 + 2a\cdot 5b + 2a\cdot 3c + 3\cdot 2 + 3\cdot 5b + 3\cdot 3c$$
$$= 4a + 10ab + 6ac + 6 + 15b + 9c$$

$\square$

In this example we dealt more with another important characteristic of mathematical language, the use of variables. **Variables** *are the names of intentionally unspecified objects.* For example, 3 is the name of a concrete number (in our imagined world of natural numbers), whereas x is the name of a "certain" number. Using the names of concrete numbers we can say something specifically about these numbers, such as that for the numbers 2 and 3 the commutativity of addition is valid: $2 + 3 = 3 + 2$. *Using variables we can say something about any numbers in a way that is almost as simple as for concrete numbers*, e.g. that for any two numbers a and b the commutativity of addition holds: $a + b = b + a$. And not only this. *In a way that is almost equally as simple as obtaining the truths about concrete numbers, thinking and calculating by means of variables, we obtain truths that are valid for all numbers. The reason is simple – variables are the names of intentionally unspecified objects, so when we use variables in thinking, we don't use anything specific about the objects they name, and our conclusions are general.* Thus, in the previous example we proved the universal principle of "multiplying brackets". *The mechanism of variables allows the transition in thinking away from statements about concrete numbers towards thinking with universal laws about numbers, retaining the simplicity of thinking with concrete numbers. This also holds generally, regardless of numbers: variables allow us a very significant jump in thinking and calculation from concrete (about specific objects) to abstract (about any objects).* The use of variables was a major advancement in math, I would even say a revolution: the key internal mechanism with which European mathematics, starting from the 16th century, moved forward in relation to math from previous cultures. A similar leap occurs in anyone's math education even today. Acquiring the mechanism of variables in the process of learning math, this transition in working with concrete objects $(1, 2, 3, \ldots)$ to any objects $(x, y, z, \ldots)$, is a very important jump in math education, that many fail to make in the correct way. *It is interesting that this means that the key thing in learning math is not acquiring math objects but rather acquiring a language in which we think about these objects.* People are quite eloquent when they speak about even more unusual things than numbers, for example, about science-fiction movies in which the laws of our world are completely messed up. Mathematical eloquence starts from the language rather than from knowledge about math objects. If I summarised into one sentence my teaching experience, I would say that *almost everybody who "does not understand math" in fact does not understand the language of math*. She or he has not mastered the particularities of mathematical language, meaning and the form of structures in this language, first of all symbolisation and use of variables, and this is what hinders her or him in mastering math.

> **The Meaning and the Importance of Variables in Mathematics**
>
> Variables are the names of intentionally unspecified objects.
>
> Using variables we can say something about any objects in almost as simple a way as we can for concrete objects. And not only this. In a way almost equally simple to the way we obtain truths about concrete objects, thinking and calculating by means of variables, we obtain truths that are valid for all objects.
>
> Thus, the mechanism of variables allows the transition in thinking: from statements about concrete objects to thinking with universal laws about any objects, while retaining the simplicity of thinking with concrete objects.

In this section there *"occurred"* another essential characteristic of mathematics, which is related to the previously described characteristics. Calculating the descriptions of numbers, closed or open, we obtained certain statements, for example the concrete statement that:

$$346 \cdot 3 + 345 \cdot 7 = 3453$$

and the universal statement (not a particularly significant one) that for all numbers a, b and c

$$(2a + 3) \cdot (2 + 5b + 3c) = 4a + 10ab + 6ac + 6 + 15b + 9c$$

Thus, we have obtained certain truths about numbers. The thinking (proving) with which we obtained these truths was reduced to calculation (application of formal rules).

> **The Importance of Language in Mathematics**
>
> In every area of math it is very important to develop a language in which thinking can be reduced to calculating. Thereby, we achieve greater security and efficiency in thinking, and a part of thinking can be left to the computer.

In *Circle 4* we will see that it is not possible to leave thinking entirely to the computer.

The tutorial on the book's website https://understandingmath.academ y/sagemath-materials/ describes how to calculate and simplify using the software *SageMath*. The calculation is simple: we enter the closed description into the input cell as on a calculator and by pressing the key *Evaluate* on the web page or by pressing the combination of keys *Shift + Enter* on the keyboard we get the answer. However, the definitive advantage of *SageMath* over the common calculator is that it can simplify open descriptions. That is to say, if we want *SageMath* to simplify $3a(2b+4)+3b(2a+5)$ we will enter into the input cell the *SageMath* command *expand*() (the comments begin with the symbol # and they have no influence on the evaluation):

```
var('a,b')
# We declare to the program that symbols 'a' and 'b'
# are variables.

(3*a*(2*b+4)+3*b*(2*a+5)).expand()
# Multiplication has to be written explicitly.
```

We will get the answer:

```
12*a*b + 12*a + 15*b
```

If we want a more elegant notation of the answer (but one we cannot use in further commands), we will add the command *show*:

```
var('a,b')
show((3*a*(2*b+4)+3*b*(2*a+5)).expand())
```

$$12ab + 12a + 15b$$

You can read more about the importance of language (especially of symbols and variables) for mathematics on https://en.wikipedia.org/wiki/ Mathematical_notation and https://en.wikipedia.org/wiki/History_ of_mathematical_notation.

1.13   Division

Here is one example of the application of division.

Example 1.13.1. In Mrduša Donja (a fictional village in Dalmatinska Zagora, a nonfictionally desolate part of Croatia) 50 locals came to a performance of *Hamlet*. Four spectators can sit on each bench. How many benches are needed so that all the spectators can watch the performance seated?

Solution. If we find the meaning of the operation of division with remainder clear, then we will easily describe the required number of benches. It is $50 \div 4$, if the remainder of division is 0, otherwise $(50 \div 4) + 1$ (in order to place somewhere the remainder as well). This can be expressed in a shorter way with the so-called **description per cases**:

$$\text{number of benches} = \begin{cases} 50 \div 4, \text{ if remainder is } 0 \\ (50 \div 4) + 1 \text{ otherwise} \end{cases}$$

In our case $50 \div 4 = 12$, (remainder 2) so that 13 benches are necessary.

$\square$

If the remainder of the division is 0, then division with remainder is an operation inverse to multiplication. If we take three mutually separated groups with 5 items each, and divide the total number of 15 items again into three groups, it is obvious that in each group there will be again 5 items each and the remainder will be 0.

Division of Natural Numbers

When in the division with remainder of the number a by the number b $(b \neq 0)$ the remainder is 0 $(a \bmod b = 0)$ (we say that a is **divisible** by b) then this operation is called **division** and the result $a \div b$ is also marked as $\boldsymbol{a : b}$.

I would like to reiterate. *Division with remainder* is an operation that can be performed with any two natural numbers provided that the divisor is not 0. *Division* is a different operation which can be performed with natural numbers only when the remainder is 0 and then it matches the previous operation. E.g. $7 \div 2 = 3$, and $7 : 2$ is not defined, while $8 \div 2 = 8 : 2 = 4$.

Division is an operation inverse to multiplication:

> ### Inverse Relationship between Multiplication and Division
>
> 1. $a : b = c$ *precisely when* $a = c \cdot b$
>
> 2. $(a : b) \cdot b = (a \cdot b) : b = a$

Please note that it has not been written for which a, b and c these statements are valid. It is boring to continually write this. Therefore, we will introduce an agreement for the future:

> ### Agreement on Allowed Values of Variables
>
> When it is not written for which numbers $a, b, c, \ldots$, a rule is valid, then it is considered that it is valid for all numbers for which the mentioned operations have meaning.

For instance, according to this agreement, since division (not with remainder – I remind you!) is present in the previous rules, the rules are valid for all the numbers a, b and c such that a is divisible by b. Such agreements are introduced for ease of expression, but they imply that you know about them. Otherwise, they could be misunderstood.

Example 1.13.2. It is well known that politicians like to shake hands (and be photographed smiling with children). At a reception, there were 69 politicians and, naturally, every one of them had to shake hands with everyone else. How many handshakes took place?

Solution. Every politician shook hands 68 times. Since there are 69 of them, and two politicians participate in every handshake, the total number of handshakes is $(69 \cdot 68) : 2 = 2346$. Who can say now that the politicians do not do anything!

The fact that in dividing the natural number a by the natural number b there is no remainder means that b can fit into a an integer number of times: there is a natural number n such that $a = n \cdot b$. This relation is important because it shows whether it is possible to equally distribute a items in b groups or not. So we will emphasise it once more.

Divisibility

We say that a natural number a is **divisible** by a natural number b if there exists a natural number n such that $a = n \cdot b$. We also say that b **divides** a, that b is **divisor** or **measure** of number a, and that a is **multiple** of number b.

For example, the divisors of the number 32 are:

$$1, 2, 4, 8, 16 \text{ and } 32,$$

so that these are the only possible ways of dividing 32 schoolchildren into equal groups. However, there is an infinite number of multiples of the number 32:

$$1 \cdot 32 = 32, 2 \cdot 32 = 64, 3 \cdot 32 = 96, \ldots$$

If the headmaster wants to equally distribute ice-cream to 32 schoolchildren, he has to buy either 32 or 64 or 96 or ... ice-creams. Whether two numbers are divisible or not we can always find out by performing the formal procedure of division with the remainder and by examining whether the remainder is 0 or not. We will thus check the divisibility of the number 114 with 9:

$$
\begin{array}{r}
114 : 9 = 12 \\
\underline{-9\downarrow\quad} \\
24 \\
6 \neq 0\;!
\end{array}
$$

Therefore, 114 cannot be divided by 9.

Every natural number is divisible by 1 and by itself. These divisors are called **trivial** (insignificant) **divisors** of a number. We call other divisors **non-trivial divisors**. Every non-trivial divisor of a number yields decomposition or factorisation of the number into a product of two smaller numbers. For example, 12 is divisible by 3 so that $12 = 4 \cdot 3$. The numbers that are multiplied are also called **factors** (lat. *factor – doer*) of the product. Thus, 4 and 3 are factors of the product $4 \cdot 3$. However, maybe we can further factorise these numbers: into products of smaller factors. The number 4 can thus be further factorised: $4 = 2 \cdot 2$. Thus also $12 = 2 \cdot 2 \cdot 3$. However, the number 3 cannot be factorised further in this way. Since factorisation of the numbers 0 and 1 is not interesting (why?), we divide the remaining natural numbers into two groups: those that are factorisable into a product of two smaller numbers, such as the number 4 (we call them **composite numbers**) and those that cannot be factorised in this way, like the number 3 (we call them **prime numbers**). Every factorisation of a number bigger than 1 into smaller factors must end eventually. If a number is prime then we say that it is its own factorisation. For example, 31 is a prime number, and we say that it is factorisation of itself: $31 = 31$. If a number is composite, it is a product of two smaller numbers. If they are prime numbers, we have obtained the required factorisation. If one of them is composite, we factorise it further, and so on. As we obtain smaller and smaller numbers, the division must come to an end. Then the number is factorised in a product of prime factors. If we order the factors in the product from smaller to bigger, such as in the factorisation $12 = 2 \cdot 2 \cdot 3$, we call such factorisation **ordered factorisation of a number into prime factors**. Thus, by successive factorisation, we will factorise any number bigger than 1 into prime factors. In whatever order we factorise, we should always obtain the same prime factors. Otherwise, if we were to obtain in one factorisation of a number, for example, the factor 5 and in another factorisation of that same number we were not to obtain the factor 5, that would mean that the factored number is both divisible by 5 and not divisible by 5. I hope this consideration is sufficiently convincing for you to accept the statement that every number bigger than 1 can be divided into prime factors and that this ordered factorisation is unique.

> ### Fundamental Theorem of Arithmetic
>
> Every natural number bigger than 1 has a unique ordered factorisation into prime numbers.

The bombastic name of this theorem naturally leads to the assumption that it is very important – as if it came from the headlines of *The Morning News*. We will apply it in order to resolve some basic problems. To mathematicians, however, it assists in researching the world of large numbers. These studies appear sometimes completely distant from reality – a seventh heaven for math adventurers. But sometimes there is a flash of discovery that affects reality in multiple ways. If a mathematician told you that he is dealing with the factorisation of hundred-digit numbers (try to imagine just how long the notation of such a number would be), you would probably look at him with suspicion. However, this is precisely the research which forms the basis for modern-day extremely important procedures of encrypting data in computer systems, telecommunications and on the internet. Therefore, when something is unknown to us, we should not make a conclusion too quickly about it, though we do tend to do this (in my case regarding some works of modern art).

The procedure for working out the factorisation of a number into prime factors is simple:

Procedure of Factorisation into Prime Factors

Using e.g. the procedure of division with remainder we are looking for a non-trivial divisor d_1 of the given number a. It is sufficient to look at the numbers such that $1 < d_1 \cdot d_1 \le a$ (why?). If there is no such divisor then the number is a prime number. If, however, we find such a divisor d_1 then we use it to perform the factorisation $a = d_1 \cdot d_2$ and repeat the procedure with numbers d_1 and d_2.

Example 1.13.3. Let us factorise into prime factors the numbers 1980 and 149.

Solution.

$$1980 = 198 \cdot 10 = 9 \cdot 22 \cdot 2 \cdot 5 = 3 \cdot 3 \cdot 2 \cdot 11 \cdot 2 \cdot 5$$

We will arrange the factorisation by writing prime factors from the smaller ones towards the bigger:

$$1980 = 2 \cdot 2 \cdot 3 \cdot 3 \cdot 5 \cdot 11$$

We will also use the notation of powers (powers will be dealt with in more detail in the next section) as an abbreviation for successive multiplication. For instance, instead of writing $5 \cdot 5 \cdot 5$ we will write 5^3 meaning that three fives are multiplied. In this notation our factorisation looks like this:

$$1980 = 2^2 \cdot 3^2 \cdot 5 \cdot 11$$

When factoring the number 149 it is not easy to find the non-trivial divisor. There are rules of divisibility that can help us to determine – by checking the digits rather than performing division – whether a number is divisible by another number. For instance, a number is divisible by 10 precisely when its last digit is 0, by 5 when the last digit is 0 or 5, by 2 when the last digit is an even number, and by 3 when the sum of its digits is divisible by 3 (you can read more about the rules of divisibility on `https://en.wikipedia.org/wiki/Divisibility_rule`). Using these rules we can easily determine that 149 is not divisible by 2, 3, 5 and 10. This means that it is not divisible by their multiples, e.g. 4, 6, 8, 9, 10, 12,... With the procedure of division we can determine that it is not divisible by 7, nor by 11. The next candidate is the number 13. Since $13 \cdot 13 = 169 > 149$, according to the procedure described in the previous box, we stop looking for non-trivial divisors and conclude that the number 149 is a prime number. Therefore, 149 is a factorisation of itself into prime factors: $149 = 149$. $\qquad\square$

In the software *SageMath* we can factorise e.g. the number 6661029375 by entering the command

```
show((6661029375).factor())
```

We will obtain the required factorisation:

$$3^2 \cdot 5^4 \cdot 7^2 \cdot 11 \cdot 13^3$$

The factorisation of a number into prime factors is like making its "X-ray image": we can find out a lot about the number from it. We will use this factorisation to find the greatest common divisor and the smallest common multiple of a set of numbers, since we will need this application later on. Let us take as example the numbers 6 and 8, and let us look at their divisors and multiples.

divisors of the number 6: 1, 2, 3, 6

divisors of the number 8: 1, 2, 4, 8

The common divisors are 1 and 2. Since 1 is a divisor of any number, common divisors of the set of numbers always exist. Since the common divisors are limited by the numbers whose divisors they are, they are finite in number, and there is the greatest among them. It is called **the greatest common divisor** of a set of numbers. In the previous case of numbers 6 and 8 we can see that it is the number 2.

multiples of the number 6: 6, 12, 18, 24, 30, 36, 42, 48, 54,...

multiples of the number 8: 8, 16, 24, 32, 40, 48, 56,...

Common multiples are 24, 48,.... Since the product of a set of numbers is a multiple of each of them, there exists a common multiple of the finite set of numbers, so that there is also their **smallest common multiple**. In the previous case of the numbers 6 and 8 we can see that it is the number 24.

Greatest Common Divisor and Smallest Common Multiple

Common divisors of natural numbers $a, b, c, ...$ always exist (it is at least the number 1) and they are smaller than or equal to all numbers $a, b, c, ...$, and so there exists the greatest among them, **the greatest common divisor**, which is marked by $\mathbf{gcd}(a, b, c, ...)$.

Also, there are always common multiples of a finite set of natural numbers $a, b, c, ...$ (this is at least their product $a \cdot b \cdot c \cdot ...$), so there is also the smallest among them, **the smallest (least) common multiple**, denoted by $\mathbf{lcm}(a, b, c, ...)$.

The factorisation of numbers into prime factors allows a faster search for the largest common divisor and the smallest common factor. To illustrate the procedure, we will take the numbers 1240 and 648. Their factorisation into prime factors is as follows:

$$2160 = 2^4 \cdot 3^3 \cdot 5$$

$$648 = 2^3 \cdot 3^4$$

The factorisation into prime factors of their common divisor is part of the factorisation of both numbers. Therefore, the factorisation of the greatest common divisor is precisely the largest common part of these factorisations. Thus, it is composed of the common factors of both factorisations, where every common factor is taken as many times as the fewest times it appears in one of the factorisations. For instance, for the previous two numbers the common factors are 2 and 3. The factor 2 appears in the first factorisation 4 times and in the second 3 times, so in the largest common divisor it appears 3 times. In the same way we have determined that the factor 3 appears 3 times in the largest common divisor, so that

$$\gcd(2160, 648) = 2^3 \cdot 3^3 = 216$$

On the other hand, the factorisation into prime factors of their common multiple contains factorisations of both numbers. Therefore, the factorisation of the smallest common multiple is the smallest such factorisation. Thus, it is composed of the factors of both factorisations, where every factor is taken as many times as the maximum number of times it appears in the factorisations. For instance, in the factorisation of the previous two numbers the factors 2, 3 and 5 appear. The factor 2 appears in the first factorisation 4 times and in the second 3 times, so in the smallest common multiple it appears 4 times. In the same way we determine that the factor 3 appears at the most 4 times, and the factor 5 once at the most. Therefore

$$\mathrm{lcm}(2160, 648) = 2^4 \cdot 3^4 \cdot 5 = 6480$$

This illustrates the general procedure of finding the largest common divisor and the smallest common multiple of a set of numbers.

Procedure of Finding the Greatest Common Divisor and the Smallest Common Multiple

In order to find $\gcd(a, b, c, ...)$ and $\mathrm{lcm}(a, b, c, ...)$, for the numbers $a, b, c, ...$ the following has to be performed:

1. factorise each of the numbers $a, b, c, ...$ into prime factors;

2. $\gcd(a, b, c, ...)$ will be obtained by making a product of the common factors of all factorisations where each factor is taken as many times as it appeared least in one of the factorisations;

3. $\mathrm{lcm}(a, b, c, ...)$ will be obtained by making a product of all factors from the factorisations where every factor is taken as many times as it appeared most in any of the factorisations.

Example 1.13.4. Let us find the gcd and lcm of numbers 135, 2520 and 300.

Solution. First we will factorise the numbers into prime factors (we will arrange the factorisations by writing prime factors from lower to higher):

$$135 = 3^3 \cdot 5$$

$$2520 = 2^3 \cdot 3^2 \cdot 5 \cdot 7$$

$$300 = 2^2 \cdot 3 \cdot 5^2$$

Common factors of all factorisations are 3 and 5. The factor 3 appears the fewest times in the third factorisation – once, and the factor 5 in the second – also once. Thus,

$$\gcd(135, 2520, 300) = 3 \cdot 5 = 15$$

All factors of the factorisations are 2, 3, 5 and 7. The factor 2 appears most times in the second factorisation (3 times), the factor 3 in the first factorisation (3 times), the factor 5 in the third factorisation (twice), and the factor 7 in the second factorisation (once). Thus,

$$\mathrm{lcm}(135, 2520, 300) = 2^3 \cdot 3^3 \cdot 5^2 \cdot 7 = 37\,800$$

$\square$

In the software *SageMath* the commands gcd and lcm give us the greatest common divisor and the smallest common multiple. However, it is only possible to input two numbers. Thus, if we wanted to find the gcd and lcm of the three numbers from the previous task, we would have to input:

```
gcd(gcd(135,2520),300)
lcm(lcm(135,2520),300)
```

1.14   Raising to the Power

We have seen that in the factorisation of a number into prime factors often the same factor is multiplied several times. If for no other reason than simpler notation, it is preferable to introduce a new operation. If, for example, the number 2 appears in the product three times, we can write it more briefly: $2^3 = 2 \cdot 2 \cdot 2$.

Raising to the Power

Each natural number a and each positive natural number b are assigned the number a^b which we obtain by multiplying b of numbers a:

$$a^b = \underbrace{a \cdot a \cdot \ldots a}_{b \text{ times}}$$

This operation is called **raising to the power** or **exponentiation** and the number $\boldsymbol{a^b}$ is called the **power** of numbers a and b. In standard notation of the power a^b of numbers a and b there is no symbol for this operation: it is rather highlighted by the order of notation of these numbers. The number which is "down" is called the **base**, and the number which is 'up" the **exponent** of the power. There is also another notation of this operation which is usually applied in computer science, in which it has its own sign: $\boldsymbol{a \textbf{ exp } b}$.

Thus, for example, in the power 2^3 the base is 2, and the exponent 3.

The operation of raising to the power is conveniently expanded to raising to the power of zero, in the following way:

Raising to the Power of Zero

1. $a^0 = 1$ for $a \neq 0$ (this convention ensures that the 2nd rule in the next box holds also for $n = m$).

2. 0^0 is not defined.

The following hold for raising to the power:

Basic Rules for Raising to the Power

1. $a^n \cdot a^m = a^{n+m}$

2. $a^n : a^m = a^{n-m}$

3. $a^n \cdot b^n = (a \cdot b)^n$

4. $a^n : b^n = (a : b)^n$

5. $(a^n)^m = a^{n \cdot m}$

6. $0^n = 0,\, 1^n = 1$

7. *For $a > 1$, $a^n < a^m$ precisely when $n < m$*

8. *For $n > 0$, $a^n < b^n$ precisely when $a < b$*

To avoid misunderstanding, I will repeat once more **the agreement**: *when we do not write for which numbers a, b, n and m the rules are valid, then they are valid for all those a, b, n and m for which they have meaning.* For instance, the first rule is valid for all a, n and m, except for $a = 0$ and $n = 0$ or $a = 0$ and $m = 0$.

On the basis of the description of raising to the power you can convince yourself of the accuracy of the mentioned rules. The next expression should convince you that the first rule is valid:

$$a^n \cdot a^m = \underbrace{a \cdot a \cdot \ldots a}_{n \text{ times}} \cdot \underbrace{a \cdot a \cdot \ldots a}_{m \text{ times}} = a^{n+m}$$

Example 1.14.1. Let us evaluate:

1. $2 + 2 \cdot 2^2$ 2. $2^{1203} : 2^{1200}$ 3. $\left(15^3 \cdot 21^2\right) : \left(35^2 \cdot 3^4\right)$

Solution.

1. Again, there are no brackets here that would define the order of calculation. However, in such cases we apply **agreement on the order of operations** *according to which, in an ambiguous situation, the raising to the power is done first, and then multiplication and division, and then addition and subtraction:*

$$2 + 2 \cdot 2^2 = 2 + 2 \cdot 4 = 2 + 8 = 10$$

2. If we calculated consecutively $2^{1203} : 2^{1200}$ we would get so confused by such large numbers that even a calculator could not help us. However, according to the 2nd rule for power:

$$2^{1203} : 2^{1200} = 2^{1203-1200} = 2^3 = 8$$

3. First we will factorise the bases into prime factors in order to reduce the number of bases to a minimum, then we will simplify the expression according to the rules of raising to the power, and only then will we calculate directly:

$$\left(15^3 \cdot 21^2\right) : \left(35^2 \cdot 3^4\right) = \left((3 \cdot 5)^3 \cdot (3 \cdot 7)^2\right) : \left((5 \cdot 7)^2 \cdot 3^4\right) =$$
$$\left(3^3 \cdot 5^3 \cdot 3^2 \cdot 7^2\right) : \left(5^2 \cdot 7^2 \cdot 3^4\right) = \left(3^5 \cdot 5^3 \cdot 7^2\right) : \left(3^4 \cdot 5^2 \cdot 7^2\right) =$$
$$\left(3^5 : 3^4\right) \cdot \left(5^3 : 5^2\right) \cdot \left(7^2 : 7^2\right) = 3^{5-4} \cdot 5^{3-2} \cdot 7^{2-2} = 3 \cdot 5 \cdot 1 = 15$$

$\square$

Example 1.14.2. Let us simplify:

1. $3a^2b^3 \cdot 6a^3bc$ 2. $(3a^2b^3)^2$

Solution. To remind you: we transform open descriptions (now mainly according to the laws of powers) usually in the order in which we would calculate them directly if we knew a, b and c. Try to determine at each step which rule (or several rules) has been applied

1. $3a^2b^3 \cdot 6a^3bc = 3 \cdot 6 \cdot a^2 \cdot a^3 \cdot b^3 \cdot b \cdot c = 18a^5b^4c$

    (do not forget, all are multiplied, as if it was written $3 \cdot a^2 \cdot b^3 \cdot 6 \cdot a^3 \cdot b \cdot c$)

2. $(3a^2b^3)^2 = 3^2 \cdot (a^2)^2 \cdot (b^3)^2 = 9a^4b^6$ $\square$

1.15 The Background Mathematical Objects

Our consideration of natural numbers naturally led to the consideration of various sets of numbers (e.g. the set of all even numbers), various relations among numbers (e.g. comparison of numbers) and various operations with numbers (e.g. addition of numbers). Sets, relations and operations were auxiliary means to say something about numbers. The same occurs in considering other objects. Moreover, **sets**, **relations** and **operations** (operations are also called **functions**) can themselves be the objects of mathematical consideration. Thus, we have seen that, for example, every set of natural numbers has its smallest element, that the relation of comparison is transitive, and the operation of addition is commutative.

Indeed, the connection between sets and natural numbers is a deeper one. In considering the natural numbers sets are not only an auxiliary means. Using natural numbers we measure sets, with relations among natural numbers we express relations among sets, and operations with natural numbers rest on adequate operations with sets. In *Circle 3* we will see that without sets we cannot even describe the world of natural numbers adequately.

Since sets, relations, and operations occur in every mathematical consideration, they are the basic mathematical objects. Therefore, we will pay special attention to them at the next levels. At this level (*Circle 1*) they will not be a special subject of observation, but we will use them instead in considering the numbers. Thus, we will develop a certain feeling about them that will be the base for the next levels. For the moment, it is important that we understand that they are also, in a certain way, imagined objects, and that as such they are equal to all other mathematical objects. Their specific characteristic is that they are comfortable auxiliary means in all considerations, even of themselves, as we will see in *Circle 4*.

1.16   Problems and Solutions

Counting

1. Is there a biggest natural number? If not, how is this possible when the universe is considered to be finite?

Decimal Notation

2. Explain why in counting, 399 is followed by 400.

3. If people had only four digits on each hand, when counting they would probably group eight (not ten) smaller units into a bigger one. How many digits would they need in such a system to record the numbers? What would be the notation for number ten in such a base?

Comparison

4. Order the numbers from the smallest to the greatest: 14 567, 14 576, 14 665, 14 656, 14 566.

5. A human can have at most 300 000 individual hairs. If the Osijek-Baranja County (in the eastern part of Croatia) has 330 000 inhabitants, can you determine that there are at least two persons with the same number of hairs, without travelling all around the county (and without disturbing bald people)?

Addition

6. Estimate the following sums and using the software *SageMath* check the estimate:

   (a) $1236 + 328$ (b) $2\,345\,235 + 2\,345\,720$ (c) $298\,356 + 278\,645$

7. In the amateur cultural art society *Autumn is Coming my Darling*, 24 members are saving for their retirement with the insurance company *Grombach & Sons* and 24 with another, *Krombach & Sons*. 10 of them, fearing that one fund might collapse, save in both funds. If there are 39 members in the society, how many do not save in either of the funds?

8. In the amateur cultural art society *Cheerful Tambourine* 20 members sing in the choir, and 12 members are engaged in naïve painting. If there are 32 members in the society, of whom everybody is engaged in something, is there a member who both sings and paints?

9. Replace asterisks with digits so that the addition is correct:

$$\begin{array}{r} 7*2* \\ +\ \ *4*2 \\ \hline 13191 \end{array}$$

10. Replace the same letters with the same digits so that the addition is correct:

$$\begin{array}{r} A \\ AB \\ +\ ABC \\ \hline BAB \end{array}$$

Proving

11. Grandpa Janko listened with interest to a report on the local radio station about the final order of teams in the regional football league. But as he is a little deaf he caught only the following information:

    (a) There were 6 teams in the competition.

    (b) In spite of expectations *Tough Guys FC* are not the champions.

    (c) In the last round *Revellers FC* overtook *Cooperative FC*.

    (d) *Tambourine FC* occupied third place.

    (e) *Deratisation FC* was placed before *Revellers FC*, but did not overtake *Tough Guys FC*.

    However, he did not hear anything about his favourite team, *Homemade Sausage FC*. Help grandpa to work out the final ranking of the teams.

12. At the end of the birthday party the solicitous host Tomo decided to check whether the drivers present were sober enough to drive home. Since he did not have a breathalyser he thought of the following test. He placed on a table in the adjacent room 5 vessels of different shapes and poured into them 5 different types of drinks. He told the drivers that a bottle, a glass, a can, a pot and a jug were placed in a certain order and that beer, brandy, juice, wine and water were in them. He also told them:

    (a) The can does not contain an alcoholic beverage.

    (b) Water and wine are not in adjacent vessels.

    (c) The vessel with beer is exactly between the pot and the vessel with juice.

    (d) The vessel with brandy is immediately to the right of the glass.

    (e) The jug and the vessel with wine are neighbours.

    (f) The can and the pot have a single neighbour each.

    He decided to let drive only those who deduced in which order the vessels were placed and what kind of drink was in which vessel. Try to pass this breathalyser test.

Basic Properties of Comparison

13. Are the following statements true?

    (a) For any natural number there is a preceding natural number.

    (b) If $a < b$ and $b < c$ then $c < a$.

    (c) In every set of natural numbers there is one natural number that is the greatest.

14. Is there a country with the smallest number of inhabitants? If there is, how can we know that without looking it up in the geographic data, which are tricky anyway?

15. Is there a country with the most inhabitants? How could we prove this using the basic properties of comparison of natural numbers?

16. A mathematician claimed that all natural numbers are interesting. For a mathematician this is far from strange. However, this one offered also the proof that it is so. Here is his proof. One property of comparing natural numbers says that in every non-empty set of natural numbers there is a number which is the smallest among them. If there were uninteresting natural numbers, then according to the mentioned property there would also exist the first uninteresting number. However, it is precisely because of this that this number would be interesting! And this is a contradiction. Therefore, there are no uninteresting numbers. What is wrong with this proof?

17. Prove out of the basic properties of the relation $<$ the following statement: *For any natural number a, not $a < a$.*

Defining

18. On the basis of the definition of relation $\leq$ and the basic properties of relation $<$ prove:

    (a) *For any a, $a \leq a$.*

    (b) *For any a and b, $a \leq b$ or $b \leq a$.*

19. Define the relation $<$ by means of relation $\leq$.

Basic Properties of Addition

20. Using the properties of addition, calculate in your head:

    (a) $1 + 357 + 17\,999$ (b) $228 + 17\,999$ (c) $899 + 1343 + 101$

21. Try to invent on your own an operation with numbers. Use the symbol $\circ$ (small circle) for this operation, which transforms a pair of numbers into one number. Try to determine at least a certain property of this operation.

22. Prove that $a + c \leq b + c$ precisely when $a \leq b$.

Subtraction

23. Estimate and with the software *SageMath* verify the estimate:

    (a) $12\,000 - 4990$ (b) $12\,345 - 8891$ (c) $345\,623 - 54\,931$

24. Replace the asterisks with digits, keeping the procedure correct:

    (a)
$$
\begin{array}{r}
9*74* \\
-\quad 6*23 \\
\hline
*11*5
\end{array}
$$

    (b)
$$
\begin{array}{r}
2 \\
-\quad 3*7 \\
\hline
34
\end{array}
$$

25. Is $a - b = b - a$ (is subtraction a commutative operation)?

Multiplication and Division

26. Estimate and with the software *SageMath* verify the estimate:

    (a) $32 \cdot 48$ (b) $2300 \cdot 21$ (c) $82345 \cdot 345$

27. Calculate yourself or using the software *SageMath*:

    (a) $23 \div 5$ and $23 \bmod 5$ (b) $1234 \div 42$ and $1234 \bmod 42$

28. Replace the asterisks with digits keeping the procedure correct:

    (a)

    $$
    \begin{array}{r}
    * * \cdot 6 \\
    \hline
    7\,8
    \end{array}
    $$

    (b)

    $$
    \begin{array}{r}
    * 2 \cdot 3 * \\
    \hline
    1\,2\,* \\
    +\quad 1\,6\,8 \\
    \hline
    1\,*\,2\,8
    \end{array}
    $$

    (c)

    $$
    \begin{array}{r}
    1\,2\,* : 8 = *5 \\
    -\quad * \\
    \hline
    4\,5 \\
    -\,4\,* \\
    \hline
    5
    \end{array}
    $$

    (d)

    $$
    \begin{array}{r}
    1\,*\,8 : 11 = ** \\
    -\,1\,1 \\
    \hline
    3\,8 \\
    -\,** \\
    \hline
    *
    \end{array}
    $$

29. We claim that $1 \cdot 5 = 5$. Why is it then that when a student gets the 5 worst grades (ones) in a certain subject, they have to take the corrective exam, instead of being included among excellent students (having the grade 5)?

30. Prove that $(a \bmod b) \div b = 0$.

Multiplication

31. Using the properties of multiplication calculate in your head:

   (a) $8 \cdot 37 \cdot 125$ (b) $1001 \cdot 20$

32. At a school dance there are 11 girls and 22 boys. One boy and one girl should start the dance. How many possibilities are available for this choice?

33. Uncle Johnny bought his nephews 4 ice-creams of different kinds (*Snowman*, *Smiley*, *Wintry* and *Jiggler* on a stick, if you really need to know), one each. In how many ways can he distribute them?

34. One of the oldest preserved mathematical texts is the so-called Rhind Papyrus from the time of ancient Egypt. It contains also the following problem. There are 7 buildings on an estate. There are 7 cats living in each building. Every cat has eaten 7 mice. Every mouse would have eaten 7 grains of wheat. Each planted grain of wheat would have yielded 7 measures of grain. How many measures of grain have been saved? How many objects have been mentioned in total?

35. Convince yourself, convince others or prove that for multiplication the associativity law $a \cdot (b \cdot c) = (a \cdot b) \cdot c$ holds.

Calculating and Simplifying

36. Calculate:

    (a) $2 + 3 \cdot (3 - 1)$ (b) $2 + 3 \cdot (2 + 3 \cdot (2 + 3))$ (c) $233 \cdot 24 - 233 \cdot 23$

37. Simplify:

    (a) $a + b + 0 + c + a + 2b + 3c$ (b) $2 \cdot (a + 2b) + 3 \cdot (2a + b)$

38. The pirates from the good ship *Rusty Hook* are not doing particularly well lately. Fleeing from the financial police they have hidden in a bay near *Our Little Town*, and they spend their days in the *Dirty Business* inn. However, the wine is bad, and the owner charges a lot: 80 kuna per litre. For food, he charges 500 kuna per pirate per day. He also takes daily 2000 kuna, allegedly in order to bribe some important people to not give them away. How many days can the pirates continue like this, if there are 7 of them, each drinks 6 litres of wine every day, and in the pirate safe they have 100 000 kuna left?

39. A company *Let Us All Go to the Cinema* owns three cinemas in which there are 5 rows with 10 seats each and two cinemas with 8 rows with 12 seats each. On Saturday and Sunday there are three screenings of the latest hit, the film *Calmly flows the Stream under the Dreamy Crown of the Willow Tree*. One ticket is 30 kuna. If they are convinced that all seats for all the shows will be occupied, how much gross income do they expect from the ticket sales?

40. During the film screening, Renato ate two large portions of pop-corns at 15 kuna each, three chocolates (7 kuna each), a bag of candies for 12 kuna and inadvertently two chewing gums at 1 kuna each (the film was very tense). He also drank two juices: 6 kuna each. Under the influence of commercials before the film, immediately afterwards he went to McDonald's and ate a hamburger for 14 kuna. If he paid 30 kuna for his ticket, how much did the visit to the cinema cost him?

41. By how much is the sum of all even natural numbers not bigger than 1000, greater than the sum of all odd natural numbers not bigger than 1000?

Division

42. Are the following divisible?

    (a) 4 by 2   (b) 2 by 4   (c) 111 by 111   (d) 112 by 7

43. Using the software *SageMath* check whether the following are divisible:

    (a) 23 503 by 23   (b) 36 738 by 254

44. After the derby between *Bumblebees FC* and *Hornets FC* which finished with the score 0:0, each fan group was dissatisfied with the behaviour of the other. In order to prevent the heated exchange of views, the police detained the most ardent fans: 127 from one team and 1 (super-ardent) from the second team. Help the police commander: how many paddy wagons will be required to drive the detained fans to the police station, given that each vehicle can accommodate 6 fans?

45. Are the following statements true?

    (a) If the number is divisible by 2, the last digit is 2.
    (b) If the last digit of the number is 0, the number is divisible by 5.
    (c) If the last digit of the number is 3, the number is divisible by 3.
    (d) If the sum of the digits of the number is divisible by 3 then the number is divisible by 3.

46. Jerry is 20 feet away from his shelter. Tom is 5 jumps away from Jerry. When Tom jumps, Jerry makes 3 steps, but one of Tom's jumps is the size of 10 of Jerry's steps. Will Tom catch Jerry?

47. Factorise into prime factors and check the results using the software *SageMath*:

    (a) 24   (b) 180   (c) 576   (d) 2808

48. Find the largest common divisor and the smallest common multiple of the following numbers and check the result by means of the software *SageMath*:

    (a) 12 and 16   (b) 6, 8 and 12   (c) 132 and 90   (d) 21, 60 and 210

Raising to the Power

49. Calculate:

    (a) 3^4 (b) 4^3 (c) 1^{10} (d) 10^1 (e) 10^0

50. Calculate:

    (a) $2 \cdot 3^{2+1}$ (b) $2^{2 \cdot 2+1}$ (c) $2^{(2^3)}$

51. Simplify:

    (a) $2x^2 y^2 \cdot 3xy^2 \cdot 4x$ (b) $\left(2x^2 y^3\right)^2$ (c) $2(xy)^2 \cdot (3x)^3$

52. Prove:

    (a) $(a^n)^m = a^{n \cdot m}$ (b) $a^n \cdot b^n = (a \cdot b)^n$

Various Problems

53. Fill out the crossword puzzle in the picture (possibly using the software *SageMath*):

Across: 1. $38+45$, 3. $72:9$, 4. $23 \cdot 18$, 7. $480:12$, 8. $84:7$, 10. $2 \cdot 2 \cdot 2 \cdot 2 \cdot 3$, 11. $526 \cdot 10$, 13. $93 \cdot 50$ 15. $675 + 428 + 325$, 17. $726 + 483 + 198$, 20. 2^5, 22. $1008:36$, 23. $13 \cdot 3$, 25. $4092:6$, 26. $3 \cdot 23$,

Down: 1. $112 - 28$, 2. $48 + 135 + 72 + 50$, 3. $24 \cdot 34$, 5. $10 + 4$, 6. $242 \cdot 20$, 9. $715 - 511$, 12. $112 + 80 + 17 + 13$, 14. $5 \cdot 8 \cdot 16$, 15. $700 + 336$, 16. $1088 - 276$, 18. $8 \cdot 6 \cdot 10$, 19. $23 \cdot 32$, 21. $7 + 7 + 7 + 7$, 24. The largest two-digit number.

54. Calculate:

   (a) $1+2\cdot3+4\cdot5\cdot6$ (b) $1+2\cdot(3+4\cdot5)+6$ (c) $1+2^{3+4}$

55. Calculate in the simplest way possible:

   (a) $12+23+34+26+37+48$ (b) $6\cdot125\cdot2\cdot14$ (c) $234\cdot7+225\cdot7$
   (d) $228\cdot7+225\cdot13$ (e) $(238+429)\cdot0$ (f) $[(2^3)^2\cdot(2^2)^4]:2^{10}$

56. Simplify the following descriptions:

   (a) $2ab+3ac+2ab+ac$ (b) $a+a^2+2a+3a^2b+3a^2+4a^2b$
   (c) $3a\cdot(2+b)+2\cdot(3+ab)$ (d) $(a^2)^3\cdot(3a^2)^2\cdot(2a)^3$

Solutions

1. There is no biggest natural number because after each one there is
   the next. Naturally, the problem is the way in which natural numbers
   exist. I understand them as an imaginary world, and this does not con-
   tradict the idea of the finiteness of the real universe. However, those
   who consider them as part of reality need to think carefully about how
   this could be brought into agreement with the idea of the finiteness of
   the universe.

2. The adding of a new one after 399 yields the presence of 10 units that
   are grouped into 1 ten so that there remain 0 non-grouped units. But
   now we have 10 tens that we group into a new hundred and here there
   are no more non-grouped tens, so that in the end we have only 4 hundreds, i.e. the number 400.

3. For the base of 8 we would need 8 digits, from 0 to 7, and the notation
   of the number ten would be 12 (1 eight and 2 ones).

4. $14\,566$, $14\,567$, $14\,576$, $14\,656$, $14\,665$.

5. If all the citizens of the county had a different number of hairs on their
   heads, then there would be as many of these numbers as there are
   citizens: 330 000. However, then at least one of these numbers would
   be bigger than 300 000. This is impossible according to the above-
   mentioned biological fact.

6. (a) 1564   (b) 4690 955   (c) 577 001

7. One member (Jolly Reveller, double bass player).

8. No

9.
$$\begin{array}{r} 7\,7\,2\,9 \\ +\ 5\,4\,6\,2 \\ \hline 1\,3\,1\,9\,1 \end{array}$$

10. $A = 8$, $B = 9$ and $C = 2$ (in discovering the digits maybe it is easier to first analyse the hundreds, rather than the ones).

11. We can easily read from (c) and (e) that *Tough Guys FC* are before *Deratisation FC*, these before *Revellers FC*, and these before *Cooperative FC*. They occupy in order places 2, 4, 5 and 6, since according to (b) *Tough Guys FC* are not champions, and according to (d) *Tambourine FC* is in third place. The remaining team, *Homemade Sausage FC*, is in the remaining first position. Thus, the final ranking looks like this: 1. *Homemade Sausage*, 2. *Tough Guys*, 3. *Tambourine*, 4. *Deratisation*, 5. *Revellers*, 6. *Cooperative*.

12. The last statement (f) says that the can and the pot are at the ends of the series. Thus, there are two options: (i) the can is in the first place and the pot in the fifth, (ii) the pot is in the first and the can in the fifth place. Let us consider the former case. Then from statement (c) it follows that the vessel with the juice is in the third place, and the vessel with beer in the fourth place. Since according to (a) the can contains non-alcoholic beverage, it follows from the previous information that this has to be water. However, water and wine according to statement (b) are not neighbours, so that according to the previous information, wine has to be in the pot. Thus, for the moment we know the following:

The remaining brandy has to be in the second place. However, this is impossible because brandy is then right from the can, and not from the glass, as claimed in statement (d). Thus, case (i) is eliminated and case (ii) remains. Now we know that the pot is in the first place, and the can is in the fifth place. According to statement (c) the vessel with the beer is in the second place, and the vessel with the juice in the third place. According to statement (a) it follows from what has previously been said that the can contains water, and according to statement (b) it follows from what has previously been said that the pot contains wine. For the moment we know the following:

The remaining fourth place must be occupied by brandy. According to statement (d) the glass is in the third place, and according to statement (e) the jug is in the second place. Thus, the bottle should be in the remaining fourth place. So the final order of the series is:

I am not sure whether the guests managed to solve the problem successfully, but Tomo certainly managed to sober them up.

13. (a) No   (b) No   (c) No (e.g. the set of all even natural numbers has no biggest member).

14. If you look at the totals of citizens of countries, among them there is the smallest number, because one of the basic properties of natural numbers is that in every non-empty set of natural numbers there exists also the smallest among them. Thus, the properties of numbers give an answer without peeping into unreliable data. Of course, this answer is incomplete, because it does not tell us which country it is. But how come that numbers can tell us anything about the reality at all? For someone who considers numbers real as well, there is no problem here. However, those who consider numbers only as imagined objects – as I do – have to explain this. My explanation is as follows. We

have connected the imagined items (numbers) with real stuff (countries). That we can assign one number (the number of its inhabitants) to each country is an assumption about the real, and no longer about the imagined mathematical world. No matter how premeditated this hypothesis is, it is precisely the one that allows us to use numbers to know and control reality.

15. Since the number of countries is finite, so the assigned number of citizens is finite as well, and obviously there exists the largest number among them. If I wanted to replace the word "obviously" with a proof, I should prove that in every finite set of numbers there exists the biggest number. The proof is as follows. Since the natural numbers are infinite, there exist numbers larger than any number from a mentioned set of numbers, and among them is also the smallest one (according to the basic property of existence of the smallest element of the non-empty set). Let us call this number n. Then its precedent, number $n - 1$, belongs to the mentioned finite set and it is the biggest number in that set.

16. The notion of an interesting number is not a precise notion. Whereas we have precise definitions to describe which number is even and which is not, we have no precise description of which number is interesting and which is not. It is precisely the ambiguity of this notion that has been used in the proof. First it has been assumed that we had a certain notion about what an interesting number is, and later we changed our minds and pronounced the first uninteresting number to be interesting.

17. The property of comparability of natural numbers says that for the numbers a and b exactly one of the following statements is true: either $a < b$ or $b < a$ or $a = b$. Since $a = a$, by the property of comparability it cannot be that $a < a$.

18. (a) According the definition of the relation $\leq$, $a \leq b$ precisely when $a < b$ or $a = b$. Since $a = a$ then according to this definition is also $a \leq a$.

   (b) Let it not be $a \leq b$. We need to show that it is then $b \leq a$. That it is not $a \leq b$, according to the definition of this relation means that neither is $a < b$ nor $a = b$. However, regarding the comparability of the relation $<$ this means that it is $b < a$. But then also $b \leq a$ (according to the definition of the relation $\leq$).

19. $a < b$ precisely when $a \le b$ and $a \ne b$.

20. (a) 18 357   (b) 18 227   (c) 2343

21. Here is one of the most uninteresting of such operations: $a \circ b = 0$. It assigns to each pair of numbers one and the same number – zero. It is easy to see that the operation has many good properties, e.g. commutativity and associativity. However, it is totally uninteresting. In math only those operations and relations survive which are in some sense interesting or useful.

22. If $a + c \le b + c$ then $a + c < b + c$ or $a + c = b + c$. However, then $a < b$ or $a = b$. Thus, $a \le b$. Analogously, the reverse is also proved.

23. (a) 7010   (b) 3454   (c) 290 692

24. (a) $\begin{array}{r} 9\,7\,7\,4\,8 \\ -\quad 6\,6\,2\,3 \\ \hline 9\,1\,1\,2\,5 \end{array}$ (b) $\begin{array}{r} 4\,2\,1 \\ -\,3\,8\,7 \\ \hline 3\,4 \end{array}$

25. No. For $a > b$, $a - b$ is a natural number, whereas $b - a$ is not defined at all in the set of natural numbers.

26. (a) 1536   (b) 48300   (c) 28 409 025

27. (a) 4 and 3   (b) 29 and 16

28. (a) $\begin{array}{r} 1\,3 \cdot 6 \\ \hline 7\,8 \end{array}$ (b) $\begin{array}{r} 4\,2 \cdot 3\,4 \\ \hline 1\,2\,6 \\ +\quad 1\,6\,8 \\ \hline 1\,4\,2\,8 \end{array}$ (c) $\begin{array}{r} 1\,2\,5 : 8 = 15 \\ -\ \ 8 \\ \hline 4\,5 \\ -\,4\,0 \\ \hline 5 \end{array}$ (d) $\begin{array}{r} 1\,4\,8 : 11 = 13 \\ -\,1\,1 \\ \hline 3\,8 \\ -\,3\,3 \\ \hline 5 \end{array}$

29. Getting a grade 1 5 times does not mean to get $1 \cdot 5 = 5$. This is where everyday speech deceives us. We read the number $1 \cdot 5$ as "one times five", but in the sense of one adding five times, and not writing down one into the grade book five times.

30. $a \bmod b$ is a number smaller than b, so that b can fit into it 0 times.

31. (a) 37 000   (b) 20 020

32. Every girl could start the dance with one of 22 boys. Thus, there are 22 possibilities of starting a dance for every selected girl. Since there are 11 girls, the total number of possibilities is therefore $22 \cdot 11 = 242$.

33. Uncle Johnny can distribute *Snowman* to any (in principle) of 4 nephews (4 ways), *Smiley* to any of the 3 remaining nephews (3 ways), *Wintry* to any of the remaining 2 nephews (2 ways) and *Jiggler* to the remaining nephew (1 possibility). Thus, there exist $4 \cdot 3 \cdot 2 \cdot 1 = 24$ ways of distributing the ice-creams.

34. There are 7 houses, there are $7 \cdot 7$ cats, they ate $7 \cdot 7 \cdot 7$ mice that would have eaten $7 \cdot 7 \cdot 7 \cdot 7$ grains of wheat which would have yielded $7 \cdot 7 \cdot 7 \cdot 7 \cdot 7 = 16\,807$ units of wheat. Overall $7 + 7^2 + 7^3 + 7^4 + 7^5 = 19\,607$ items have been mentioned.

35. $a \cdot (b \cdot c) = a + a + \ldots + a$ ($b \cdot c$ times is added a) $= a + a + \ldots + a$ (b times) $+a + a + \ldots + a$ (b times) $+ \ldots$ (c times are these groups added with b sums of number a in each group) $= a \cdot b + a \cdot b + \ldots + a \cdot b$ (c times is added $a \cdot b$) $= (a \cdot b) \cdot c$.

36. (a) 8   (b) 53   (c) 233

37. (a) $2a + 3b + 4c$ (b) $8a + 7b$

38. Let us calculate the daily costs of the pirates. Every one of them spends $6 \cdot 80 = 480$ kuna on wine and further 500 kuna on food. This is a total of 980 kuna. Since there are 7 of them, they spend on this a total of $980 \cdot 7 = 6860$ kuna. One should add the 2000 kuna *bribe*, so their total daily costs are 8860 kuna. They will be able to keep up this tempo $100\,000 \div 8860 = 11$ days. On the twelfth day they will have only $100\,000 \bmod 8860 = 2540$ kuna and they will have to set sail or rob somebody (I would suggest the owner of the inn).

39. $2 \cdot 3 \cdot (3 \cdot 5 \cdot 10 + 2 \cdot 8 \cdot 12) \cdot 30 = 61\,560$ kuna

40. 121 kuna

41. Every positive even number not bigger than 1000 has by one a smaller associated odd number. Thus, the difference between even and odd numbers equals the number of non-zero even (= odd) numbers not bigger than 1000. Since they together yield the first thousand natural numbers, thus there are 500 even numbers, and the requested difference is also 500.

42. (a) Yes (b) No (c) Yes (d) Yes

43. (a) No (b) No

44. Since $127 \div 6 = 21$, and $127 \bmod 6 = 1$, they will need 22 paddy wagons for one support group. Although they still have room in one paddy wagon, due to the "nature of the problem", they will have to order another paddy wagon just for the *Super-ardent*.

45. (a) No (b) Yes (c) No (d) Yes

46. Jerry will manage to escape from Tom. When Jerry gets to the hole, Tom will be more than three of Jerry's steps away from it.

47. (a) $2^3 \cdot 3$ (b) $2^2 \cdot 3^2 \cdot 5$ (c) $2^6 \cdot 3^2$ (d) $2^3 \cdot 3^3 \cdot 13$

48. (a) 4 and 48 (b) 2 and 24 (c) 6 and 1980 (d) 3 and 420

49. (a) 81 (b) 64 (c) 1 (d) 10 (e) 1

50. (a) 54 (b) 32 (c) 256

51. (a) $24x^4y^4$ (b) $4x^4y^6$ (c) $54x^5y^2$

52. (a) $(a^n)^m$ means that we multiply m of numbers a^n. But a^n means that we multiply n of numbers a. Thus, we multiply in total $n \cdot m$ of numbers a, so that the value of this expression is $a^{n \cdot m}$ (b) analogue (a)

53.

8	3		8		4	1	4
4	0		1	2		4	8
	5	2	6	0			4
		2		4	6	5	0
1	4	2	8		4		
0			1	4	0	7	
3	2		2	8		3	9
6	8	2		0		6	9

54. (a) 127   (b) 53   (c) 129

55. (a) 180 (we have three sums of 60 each)

   (b) 21000 (we break up the product into even more factors and multiply first 2 and 5)

   (c) 1820 (we factor out the common factor 7)

   (d) $228{\cdot}7+225{\cdot}13 = (3+225){\cdot}7+225{\cdot}13 = 21+225{\cdot}(7+13) = 21+4500 = 4521$

   (e) 0 (multiplying with zero yields zero)

   (f) $[(2^3)^2 \cdot (2^2)^4] : 2^{10} = [2^6 \cdot 2^8] : 2^{10} = 2^{14} : 2^{10} = 2^4 = 16$

56. (a) $4ab + 4ac$ (b) $3a + 4a^2 + 7a^2 b$ (c) $6a + 5ab + 6$ (d) $72a^{13}$

2

Integers
(or what's on the other side of zero)

In this chapter we will recall what we know about integers:

* how we imagine them and what their purpose is;

* why and how we compare, add, subtract, multiply and divide them;

* what the properties of these operations with integers are and why they are better than the properties of these operations with natural numbers;

* how we calculate and simplify expressions using the rules for signs.

We will talk also about how good *Our Bank* is, how to use a current account, why caution is needed with negative numbers and that money is not everything. We will get to know Romeo and Juliet, in the exercise problems we will meet Terrible Joe, take a ride in an elevator in the *Better Life* mine and much more.

2.1 Oriented Counting

It is not the same if you win a match by 3 goals or you lose by 3 goals, if you have 50 kuna or if you owe 50 kuna, whether the temperature is 15 degrees above zero or 15 degrees below zero.... These are examples of quantities for which it is not enough just to know their amount – we must also determine in which of two possible states or orientations they are. We have to know whether they refer to gain or loss, temperature above or below zero, and so on. Such quantities are called **oriented quantities**. In order to measure them it is not sufficient to have just natural numbers – tools for measuring orientation are also necessary. One orientation can be measured by positive natural numbers. Usually the orientation which we *"prefer"* is selected, like the score difference in winning, or positive gain, but basically this is an arbitrary choice. Such orientation is called **positive orientation**. When we say that the temperature is 15 degrees, this means that it is 15 degrees above zero. However, for the other, so-called **negative orientation**, for example the score difference in losing, we must have new measures: new numbers. We will introduce (imagine) new numbers by creating from every positive natural number n its *"clone"*: a new number that will be denoted with $-n$. These numbers will be used to measure negative orientations. When we say that the temperature is -15 degrees, it means that it is 15 degrees below zero and that we have to wear warm clothes. Now we have measures for all quantities in both orientations. The fact that I have 50 kuna is measured by the number 50, and the fact that I owe 30 kuna is measured by the number -30. Also the score difference 1 is a reason for celebration, and the score difference of -1 means that the blame is on the referee.

As an example for the oriented quantity we will use the cash balance on a current account. Namely, the bank *Our Bank* promised me to forgive all my debts if I promote it by advertising. Thus, fortunately, we have *Our Bank* which will help us understand what the oriented values are. On the current account (of course, at *Our Bank*) you can have, for example, 500 kuna. The balance then on your account is 500. But you can also borrow money from the bank, so that you owe, say, 1000 kuna. Then the cash balance of your account is -1000. However, there is also one

border state. It is the state when you have no money on your current account, but neither do you owe anything. Since this is the beginning of counting in both directions (in the direction of having and in the direction of owing), for the measure of this state we will take the already known number 0.

Integers

The measures of oriented quantities are called **integers**. A set of all integers is marked by $\mathbb{Z}$. It contains zero (0), the positive natural numbers (1, 2, 3, ...) that are also called **positive integers**, and their clones $(-1, -2, -3, \ldots)$, that are called **negative integers**.

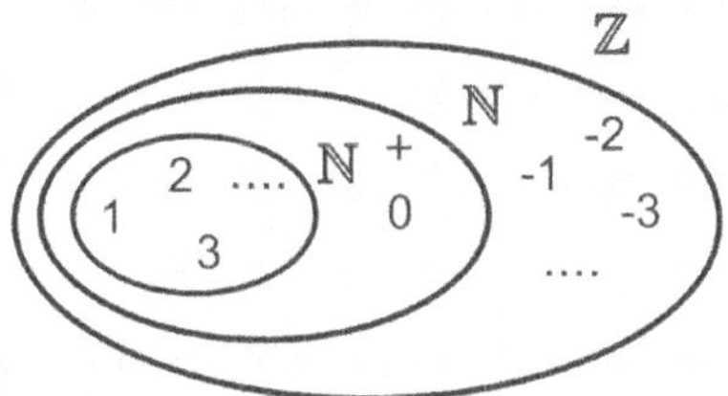

Every integer contains information about orientation and amount. Negative numbers are measures of negatively oriented quantities and they have the **sign − (minus)**. Positive numbers are measures of positively oriented quantities and they have the **sign + (plus)**. In order to emphasise this, sometimes instead of, say, 3 we write +3. The number 0 is a measure of nothing, so it is not assigned a sign (sometimes it is said to have both signs: $+0 = -0 = 0$). Regarding amount, -3 and 3 measure the same amount. They indicate that there is 3 of something, regardless of whether it refers to a loss or a gain. This amount is measured by a natural number that we call **amount** or **absolute value of the integer**. For integer a the amount of a is denoted $|a|$:

$$|-3| = 3, \quad |3| = 3, \quad |0| = 0$$

In order to better follow what is going on with a current account, we can represent it graphically with a straight line in the following way. We will select one point on the line that will represent the state 0 and will be called the origin, and one direction that we will call **positive direction**. In this direction we will assign positive numbers consecutively to points equidistant from one another, and in the opposite direction, that we will call **negative direction**, we will assign negative numbers to the points in the same manner.

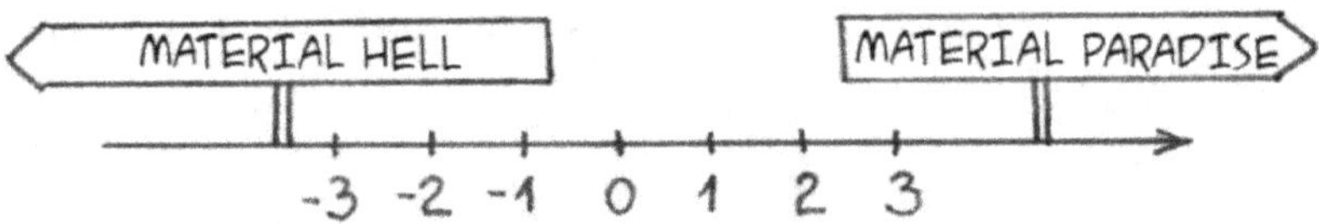

In this graphical environment, *integers are measures of the position of points on the line*. The sign of a number shows in which direction from the origin the position of the measured point is, and the amount how many steps the point is from the origin. For example, in order to arrive at point -4, we need to go from the origin 4 steps to the left:

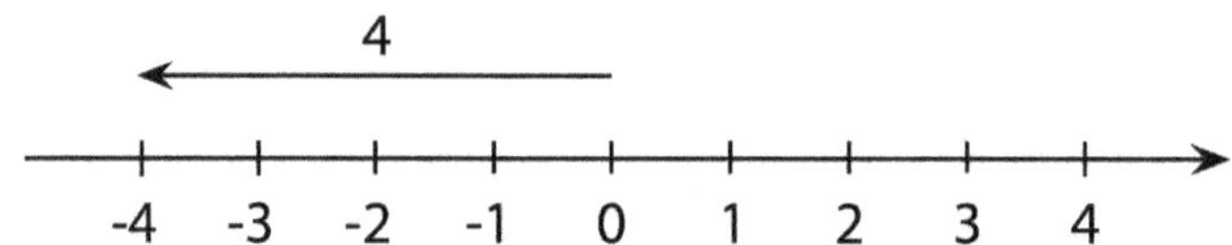

In this graphical interpretation of a current account, this progress signifies that in the beginning you had nothing, and only then did you borrow 4 kuna from the bank.

2.2 Comparison

The further to the left I go, the worse my material situation becomes; the more to the right I go, the better the situation gets.

Of course, money is not everything. There is also gold, as well as diamonds, precious stones and so on. However, looking just at the current account, if we compare two situations, whichever is to the left will always be worse.

Comparison of Integers

For integers a and b we write $a < b$, if number a is more to the left on the line than number b. A little more precise (but also more complex), is the following explanation. All negative numbers are lesser than zero, and together with zero they are lesser than all the positive numbers. If both numbers are positive, then this refers to the comparing of natural numbers. If both are negative, then the lesser one is the number that has the bigger amount. Yet more precise (but still more complex):

$a < b$ precisely when one of the following conditions is met:

1. a and b are positive and $a < b$, when they are compared as natural numbers;

2. $a = 0$ and b is positive;

3. a is negative and $b = 0$ or b is positive;

4. a and b are negative and $|a| > |b|$, when they are compared as natural numbers.

For instance, $-100 < -4$ because $|-100| > |-4|$. Naturally, this is immediately clear: it is better to owe 4 kuna than 100 kuna. However, the precise description enables a formal procedure of comparing.

Example 2.2.1. Let us order the numbers 0, -3, 50, -90 and 1000.

Solution. Negative numbers are lesser than zero, and zero is lesser than positive numbers. Thus -3 and -90 are lesser than 0, and 0 is lesser than 50 and 1000. Negative numbers are compared contrary to comparison of their amounts, and positive numbers are compared as natural numbers are. Thus we obtain:

$$-90 < -3 < 0 < 50 < 1000$$

The properties of comparison of integers are similar to the properties of comparison of natural numbers, except that there is no smallest number.

Properties of Comparison of Integers

1. *For any number a there is the first number s that is greater and the first number p that is lesser than a* (**the existence of the immediate successor and predecessor**)

2. *For any numbers a, b and c, if $a < b$ and $b < c$, then also $a < c$* (**transitivity**)

3. *For any a and b, either $a < b$ or $b < a$ or $a = b$* (**strict comparability**)

2.3 Addition

How can changes in the current account be described ? Let us say that I have 3 kuna in my account. Change can occur if money is deposited in, or withdrawn from the account. If I deposit 5 kuna in the account the amount will be increased by 5 kuna and the new balance will be 8 kuna.

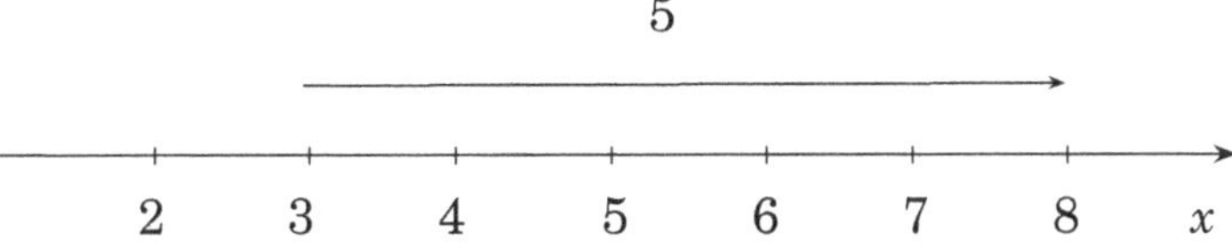

But if I withdraw 5 kuna, the balance will be reduced by 5 and the new balance will be -2:

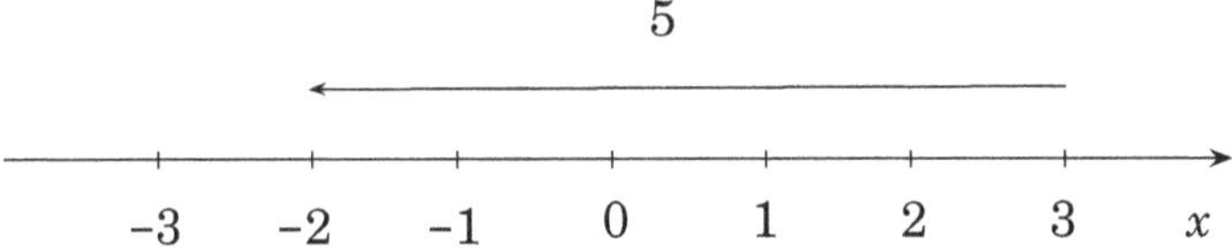

The new balance depends on the old balance and the quantity of change. The balance is an oriented quantity and we measure it by integers. But *even the change is an oriented quantity*! It may occur in two ways: by a certain amount being deposited in the account or by a certain amount being withdrawn from the account. Therefore, it will be measured by integer. Depositing money in the account will be measured by a positive, and withdrawing by a negative number. Thus, the +5 change means that we have deposited 5 kuna in the account, and the -5 change that we have withdrawn 5 kuna from the account. The new balance occurs when the change is added to the old balance. Restricted to associated numbers, the operation will be called **addition of integers**. *Concerning application, the basic meaning of the addition of integers is that it is a numerical expression for the next situation*:

Addition of Integers as Adding the Change

If integer a is the measure of the current balance and integer b is the measure of change, then $a + b$ is the measure of the new balance.

old balance + change = new balance

Thus, in the previous example we saw that $3 + 5 = 8$ and that $3 + (-5) = -2$. When we work with positive numbers the operation matches the adding of natural numbers.

In the same way we can calculate that $-3 + 5 = 2$ and that $-3 + (-5) = -8$.

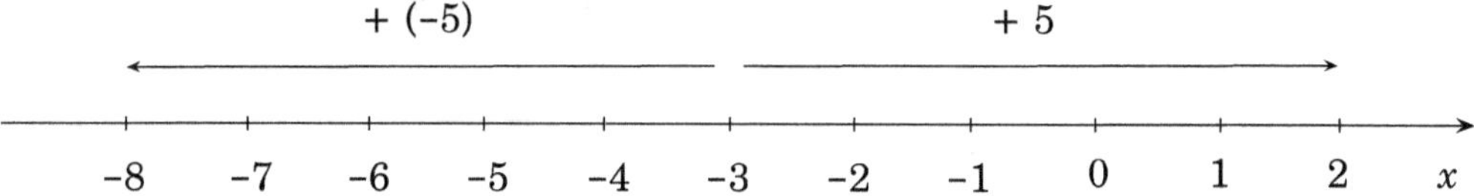

Visualised geometrically on the line, *the adding of a positive number means moving to the right for the amount of this number, and the adding of a negative number means moving to the left for the amount of this number.*

Addition of Integers on the Line

$a + b$ is an integer that we obtain when we move from number a for $|b|$ units in the direction of greater numbers, if b is positive, and in the direction of lesser numbers if it is negative. If $b = 0$ then we "*stay*" at a.

$$a + b \quad a \quad a + b$$
$$for\ b < 0 \qquad for\ b > 0$$

We can use such a description when we add small numbers. But, how will we add $-325 + 1238$? Certainly not by moving from -325 by 1238 numbers towards greater numbers. In such cases, the description of addition of integers by means of addition and subtraction of natural numbers will help us, because for these we know the formal procedures of calculation. Since the adding of zero is straightforward, the following situations remain.

1. **Addition of positive numbers**. We have already seen that it is *the same as the addition of natural numbers*. Thus, we know how to calculate:

$$325 + 1238 = 1563$$

2. **Addition of negative numbers**. How much is $-325 + (-1238)$? Since we continue moving along the negative side, from the number -325 by the amount of the number -1238, the amount of the result equals the sum of the amounts of these two numbers, and the sign of the result is negative. Thus,

$$-325 + (-1238) = -(325 + 1283) = -1563$$

It is clear that this is valid also in general. If both numbers are negative, the result will be obtained by adding their amounts and putting in front of the sum the sign '$-$':

$$a + b = -(|a| + |b|)$$

Thus we have reduced the problem once more to addition of natural numbers.

3. **Addition of numbers of different signs**. How much is $325 + (-1283)$? If we start from 325 and go leftwards, we will use 325 steps to arrive at zero, and we will continue taking $1283 - 325$ steps in the negative direction. Thus,

$$325 + (-1283) = -(1283 - 325) = -958$$

By analysing the remaining situations that occur when numbers of opposite signs are added, we can convince ourselves of the correctness of the following rule: *numbers of different signs are added by looking at their amounts: from the bigger amount the smaller is subtracted, and the result is given the sign of the number that has the bigger amount.* Thus, the calculation is reduced to subtraction of natural numbers. As we have seen, in order to obtain $325 + (-1283)$, we first look at the amounts of the numbers ($|325| = 325$, $|-1283| = 1283$), we subtract the smaller amount from the bigger amount ($1283 - 325 = 958$), and we assign to the result the sign of the number that has the bigger amount (the sign of the number -1283 is $-$). Thus the end result is -958.

The previous example show that addition can be most precisely described directly, without reference to oriented quantities and their change or, indeed, the geometrical representation.

Procedure of Adding Integers

The sum $a + b$ is calculated depending on the case. The addition of zero does not change the number. If the numbers have the same sign, we add their amounts and we assign to the result their common sign. If the numbers have different signs, we subtract the smaller from the bigger amount and we assign to the result (if different to 0) the sign of the number with the bigger amount.

1. $0 + b = b$, $a + 0 = a$

2. If $a > 0$ and $b > 0$ then $a + b$ is the sum of natural numbers

3. If $a < 0$ and $b < 0$ then $a + b = -(|a| + |b|)$

4. If a and b have different signs and let us say that $|a| > |b|$, then $a + b = p(|a| - |b|)$, where p is the sign of number a

---

Example 2.3.1. Is it warmer when the temperature decreases from 10 °C by 12 °C, or if it increases from -12 °C by 10 °C?

Solution. *Regarding the meaning of addition of integers as a numerical expression for adding changes to the current state*, we know that in the first case the new temperature is $10 + (-12) = -2$ °C, and in the second it is $(-12) + 10 = -2$ °C. Thus, the final temperatures are equal. This example shows that in addition of integers commutativity is also valid. $\square$

2.4 Opposite Number

As we have extended the comparison and addition of natural numbers to comparison and addition of integers, we will do the same with other operations. The operation that assigns the natural number n the corresponding negative number $-n$ will be extended to all integers in the following way. All integers except zero occur in pairs of numbers that have the same amount, but inverse orientation. For instance, -3 and 3 are such a pair. As the operation $-$ transforms the number 3 into that other number, -3, we will imagine that the same happens also with -3, that it is transformed into that other, the number 3.

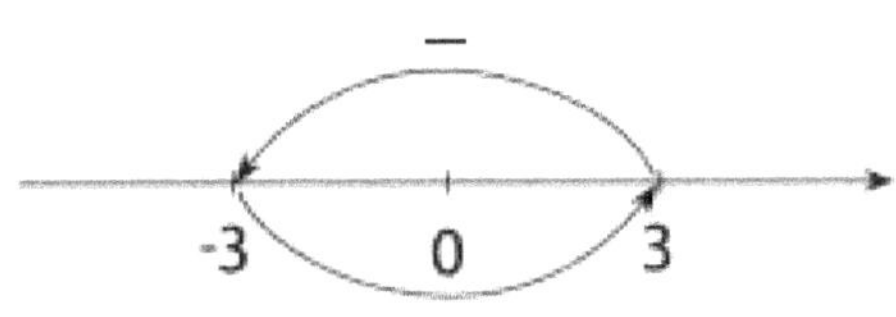

Thus, $-(-3) = 3$. In general, we will define for every negative number $-n$ (where n is a natural number) that it is:

$$-(-n) = n$$

We still need to define this operation with zero. We will accept that $-0 = 0$. Why? Now, for any integer a we know what $-a$ is.

Opposite Number

For $a \neq 0$, the number $-\boldsymbol{a}$ is the number that has the same amount as number a, but the opposite sign. For $a = 0$ we define $-0 = 0$.

This number is called the **opposite number** of number a.

Looked at in a content-wise way, the operation of taking the opposite number is an operation of orientation change. Geometrically described, this is a rotation of the number by 180°C about the origin.

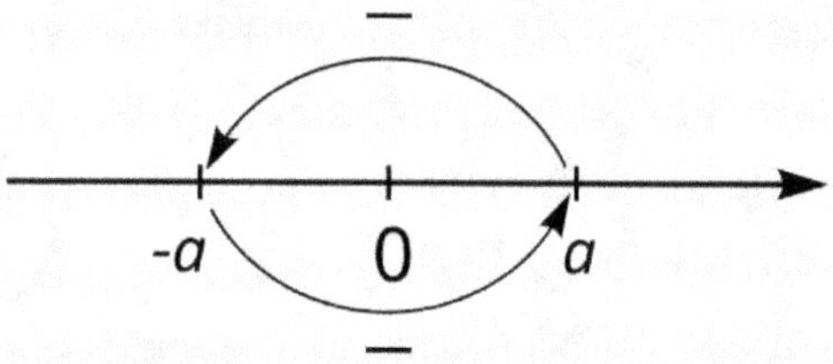

You can see from the figure (and this, of course, can be proved) that the following properties of the opposite number hold:

Basic Properties of the Opposite Number Operation

1. $|-a| = |a|$

2. $-(-a) = a$

3. $a + (-a) = 0$

2.5 Subtraction

The subtraction of natural numbers will also be extended to all integers. Addition was described as *addition* of money to the current account, in which case the adding of a negative number means withdrawing money from the account. Subtraction is intuitively understood as *withdrawing* money from the account. In this case, the subtraction of a negative number will mean withdrawing a debt from the account, or in other words, adding money to the account. If, for example, we have 3 kuna in the account, and withdraw 2 kuna, the new balance will be $3 - 2 = 1$, the same as in the case of subtraction of natural numbers. If, however, we withdraw 5 kuna, the new balance will be $3 - 5 = -2$. If we withdraw -5 kn, which is another way of saying that we add 5 kuna, the new balance will be $3 - (-5) = 3 + 5 = 8$ kn.

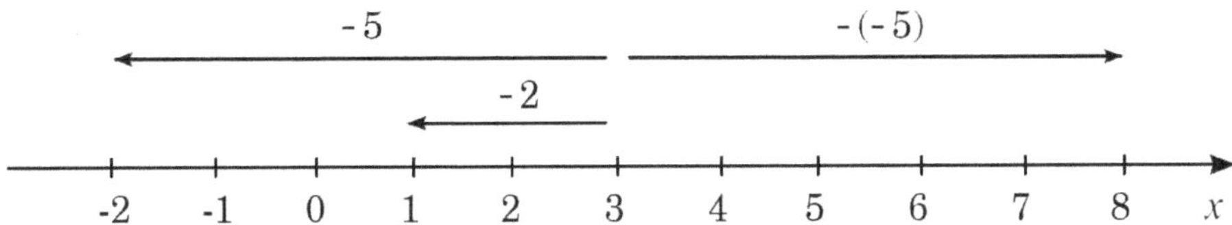

Geometrically expressed, the subtraction of a positive number means going to the left by the amount of this number, while the subtraction of a negative number means going to the right by the amount of this number.

We can see that withdrawing a certain amount from the current account is the same as the addition to the account of the opposite number of kuna. Withdrawing 5 from the current account, means adding -5, and taking -5, means adding 5. Thus, the subtracting of a number corresponds with the adding of the opposite number. Therefore, it will be defined precisely in this summarised manner:

Subtraction of Integers

$$a - b = a + (-b)$$

Example 2.5.1. If the temperature was $-8\ ^{\circ}\text{C}$, and it is now $10\ ^{\circ}\text{C}$, what is the change in temperature?

Solution. *The change is always obtained by subtracting the initial state from the end state:*

Subtraction of Integers as the Change of State

If integer a is the measure of the final state and integer b is the measure of initial state then $a - b$ is the measure of the change.

Change = final state − initial state

Subtraction is precisely the numerical expression of this relation. Thus, in this example we will obtain the amount of change there is when we subtract the initial temperature (-8) from the final temperature (10):

$$10-(-8)=10+8=18\ {}^\circ\mathrm{C}$$

$\square$

The subtraction of integers corresponds (in the set of natural numbers) to *the subtraction of natural numbers*, when this is defined. Therefore, we use the same sign ($-$) for both operations. Actually, this is *a little bit* incorrect, but convenient. Thus, $5-3$ is the same number regardless whether we consider $-$ as a sign of subtraction of integers, or subtraction of natural numbers. However, *unlike in the subtraction of natural numbers, now we can subtract any two numbers.* While in the set of natural numbers we cannot subtract the number 5 from the number 3, with integers we can do this: $3-5=-2$.

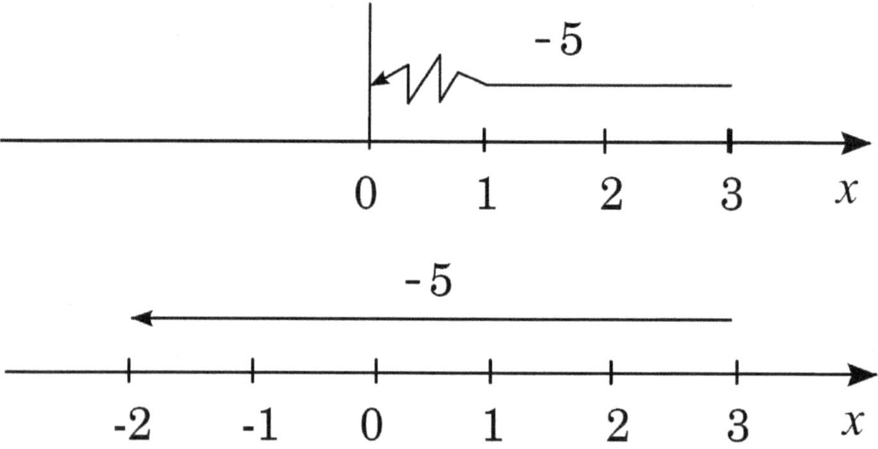

2.6 Addition and Subtraction

The set of integers has better properties than the set of natural numbers. The operations are, namely, extended operations with natural numbers. Here all the good properties of these operations are retained (which we are going to see now), and yet another new property is obtained: *subtraction is always possible*.

Basic Properties of Addition of Integers

1. $a + b = b + a$ (**commutativity of addition**)

2. $(a + b) + c = a + (b + c)$ (**associativity of addition**)

3. $a + 0 = 0 + a = 0$ (**neutrality of zero**)

4. $a - b = c$ *precisely when* $c + b = a$ (**inversion of addition and subtraction**)

5. $a + c < b + c$ *precisely when* $a < b$ (**compatibility of comparison and addition**)

Among the above-mentioned properties maybe only one thing is not completely clear, and that is the commutativity law. Why, by adding the number 5 to the number -3, do we get the same as when we add to the number 5 the number -3? To convince you this is really so, rather than using a proof from the definition of subtraction, I would like to give you the following example. Let us say that you have two current accounts: the balance of the first one is -3, and in the second one is 5. If you transfer 5 from the second account to the first one, all your money will be in the first account and it will amount to $-3 + 5$. If you want to close the first account, then you have to transfer the debt from the first account to the second one, and all your money will then be in the second account and it will amount to $5 + (-3)$. In whichever way you transfer money from one account to the other, you will be getting neither richer nor poorer. There will always be the same amount of money. Thus, $-3 + 5 = 5 + (-3)$. We could apply the same thinking to any numbers a and b. So $a + b = b + a$.

Example 2.6.1. Let us prove, by means of the basic properties of addition of integers and the basic properties of the opposite number operation, the following:

1. The opposite number of a number a is a unique number that, when added to the number a yields zero:

   If $a + x = 0$ then $x = -a$

2. The law of cancellation is valid:

   If $a + c = b + c$ then $a = b$

Solution. Try to identify at every step which property is being used:

1. $a + x = 0 \quad | +(-a) \quad \rightarrow \quad (-a) + (a + x) = (-a) + 0 \quad \rightarrow$

   $((-a) + a) + x = -a \quad \rightarrow \quad 0 + x = -a \quad \rightarrow \quad x = -a$

2. $a + c = b + c \quad | +(-c) \quad \rightarrow \quad a + c + (-c) = b + c + (-c) \quad \rightarrow$

   $a + 0 = b + 0 \quad \rightarrow \quad a = b$ $\qquad\qquad\qquad\qquad\qquad\qquad\qquad\square$

From the mentioned properties we can also prove the following important rule about the sign of the sum and difference:

Sign of Sum and Difference

$$-(a + b) = -a - b \qquad\qquad -(a - b) = -a + b$$
$$-(-a + b) = a - b \qquad\qquad -(-a - b) = a + b$$

Formally speaking, these rules are expressed uniformly in the following manner (we assume that $a = +a$):

- *If the brackets are preceded by the sign '−', we eliminate this sign and the brackets, and change the sign for all members that added yield the expression in the brackets.*

- *If the brackets are preceded by the sign '+' we simply delete the sign and the brackets.*

For instance, $-(2 - 3) = -2 + 3$, whereas $3 + (-2 + 4) = 3 - 2 + 4$.

Let us prove, by way of example, that $-(a + b) = -a - b$, that is, that $-a - b$ is the number opposite to $a + b$. According to the property of the opposite number this is the only number that added to the given number yields zero. So we need to prove that the sum of $-a - b$ and $a + b$ equals zero (try to identify which properties are used in the calculation):

$$(a + b) + (-a - b) = a + b + (-a) + (-b) = a + (-a) + b + (-b) = 0 + 0 = 0$$

We would prove similarly the remaining rules.

Well, it would be correct now to tell you something also about mathematical incorrectnesses. Although we can have different symbols for the same object, *the basic rule of accurate expression requires that a single symbol cannot denote several items, but only one, so that one will know what is being talked about.* Otherwise there may be misunderstandings, just as in everyday life. For instance, on the packaging of Vegeta (a popular Croatian condiment) you can see a smiling chef who is showing with his thumb touching the index finger the symbol of something excellent. However, if you go for instance to Greece and show someone this sign, you might end up in hospital since there this is an offensive sign (I will restrain myself from describing which offence is implied). Thus, this is a symbol that has two (and completely contradictory) meanings. How are we to avoid misunderstandings? In order to accurately determine the meaning of a symbol, you have to take care about the situation (context) in which the symbol occurs. The same happens in math as well. Instead of constantly fabricating new

symbols for new items, sometimes the same symbols are used for different items, but the context will provide the clear meaning. We have used the sign − (minus) as the symbol for subtraction of integers, but also as the symbol for the operation of the opposite number. It is clear from the context which meaning is intended. In the expression $3 - 5$ it refers to subtracting of integers since subtraction is the operation that acts on two numbers. However, in the expression $3 + (-5)$ it refers to the operation that assigns to the natural number the opposite number since it acts on one number. There will be such situations later in these books as well. This avoids the accumulation of increasingly inconvenient symbols. Moreover, sometimes this is done on purpose in order to have simpler rules. This is precisely the case with the sign −, as we will see now.

Subtraction does not support commutativity and associativity. For instance, $3 - 2 \neq 2 - 3$ and $5 - (1 - 2) \neq (5 - 1) - 2$. However, there is a simple system for calculating subtractions. According to its very definition, subtraction is addition of the opposite number, and we will consider it precisely in this manner. Just as we consider that $3 - 2$ is another expression of $3 + (-2)$, so we will consider that $3 - 5 - 7 + 10$ is just the shortened record for $3 + (-5) + (-7) + 10$. Thus, we can consider it as the addition of the numbers 3 and -5 and -7 and 10 (it is even comfortable to read this expression in this way):

$$3 - 5 - 7 + 10 = 3 + (-5) + (-7) + 10$$

This allows us to apply good properties of addition and therefore to switch members together with their signs in whatever way we wish to do:

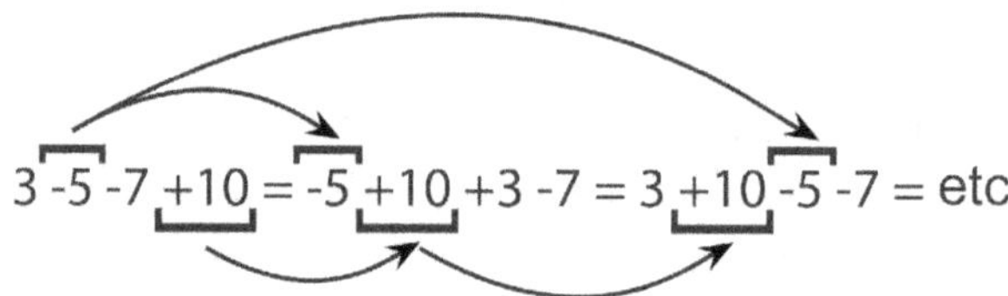

The operations + and − are considered as signs of numbers that are added. Thus the very calculation is reduced to the already described procedure of addition:

$$3 - 5 - 7 + 10 = -2 + 3 = 1$$

These simple rules are possible precisely because we (incorrectly) introduced the same symbols for operations and for the signs of numbers.

Simple Procedure of Addition and Subtraction

1. According to the rule for the signs of brackets, let us eliminate brackets.

2. Addition and subtraction are considered signs of numbers that are added to one another.

In such notation the calculation is eventually reduced to addition and subtraction of natural numbers.

Example 2.6.2. Let us calculate:
1. $12 - 23 - 27 + 52 - 8$,   2. $6 + (-2 + 3) - (-2 + 3 - 5)$

Solution.

1. Usually, all positive numbers are added and then all negative numbers, and the obtained numbers are then subtracted:

$$12 - 23 - 27 + 52 - 8 = 64 - 58 = 6$$

2. We could calculate in the order imposed by this description, calculating first what is within brackets. However, in such cases it is usually simpler to get rid of the brackets and then to add and subtract:

$$6 + (-2 + 3) - (-2 + 3 - 5) = 6 - 2 + 3 + 2 - 3 + 5 = 6 + 5 = 11$$

In the calculation we used $-2 + 2 = 0$, $3 - 3 = 0$ and in addition we can forget 0 because of the rule $x + 0 = 0 + x = 0$. $\square$

Example 2.6.3. The company *Bankruptcy* Ltd. has in its account a minus (debit) of 120 000 kuna. From sales they received 23 000 kuna on their account, and from the state budget they got financial help of 40 000 kuna. However, in the meantime, wages were raised by 90 000 kuna and the bill from the supplier has arrived: 28 000 kuna. What is the current balance of the current account? What has been the total change in the balance?

Solution. We will add all changes to the initial balance, respecting their signs:

$$-120\,000 + 23\,000 + 40\,000 + (-90\,000) + (-28\,000) =$$

$$-120\,000 + 23\,000 + 40\,000 - 90\,000 - 28\,000 = -175\,000$$

The change is always obtained by subtracting the initial balance from the final balance:

$$-175\,000 - (-120\,000) = -175\,000 + 120\,000 = -55\,000$$

The negative sign means that there has been change for the worse. $\square$

Let us emphasise once more the basic pattern of applying integers:

> **States and Changes**
>
> **old state + change = new state**
> **change = new state − old state**

2.7 Multiplication and Division

What remains to be seen is how to multiply and divide integers. In the case of natural numbers, multiplication is shortened to the addition of the same number. As we imagined then what $2 \cdot 3$ means, so now also we can imagine what $(-2) \cdot 3$ means. Simply, the number -2 is taken 3 times in addition:

$$(-2) \cdot 3 = (-2) + (-2) + (-2) = -6$$

But will we multiply with a negative number? What will $2 \cdot (-3)$ or even $(-2) \cdot (-3)$ be? The number 3 says how many times one needs to add a number, and minus says that the opposite number needs to be added. Therefore, it seems acceptable (does it?) that $2 \cdot (-3)$ means to add -2 three times:

$$2 \cdot (-3) = (-2) \cdot 3 = -6 \qquad (-2) \cdot (-3) = (-(-2)) \cdot 3 = 2 \cdot 3 = 6$$

Multiplication of Integers

Multiplying by a positive number is shortened to the addition of the same number, multiplying by zero yields zero, and multiplying by a negative number is changing the sign and then multiplying by its amount. Multiplication considered in this last manner always yields a result whose amount equals the product of the amounts of the numbers to be multiplied.

$$a \cdot b = \pm |a| \cdot |b|$$

Which sign is to be used is determined simply by **the rule for the sign of the product**:

- *if a and b have the same signs, the result is assigned the sign +;*

- *if a and b have different signs, the result is assigned the sign −.*

The rule is often presented in the following diagrammatic way:

$$+ \cdot + = +$$
$$+ \cdot - = -$$
$$- \cdot + = -$$
$$- \cdot - = +$$

It is easy to see that this rule regarding signs is valid also in general.

Sign Properties in Multiplication

1. $a \cdot (-b) = (-a) \cdot b = -a \cdot b$

2. $(-a) \cdot (-b) = a \cdot b$

Example 2.7.1. Romeo was more than happy when Juliet told him that she had tripled the balance of his current account. He only found it a little weird that she told him that immediately after having returned from shopping. If the balance in the account was -550 kuna, how much is it now? How much has the balance changed?

Solution. The new balance is $(-550) \cdot 3 = -1650$. The balance changed by $-1650 - (-550) = -1100$ kn (*change = new balance − old balance*). $\quad\square$

Multiplication of integers can also be introduced in a *structural way*: according to its role in the structure of the mathematical world formed by integers. The aim is to expand multiplication of natural numbers to all integers keeping - during this expansion - all the good properties of multiplication. For instance, if we want to keep the commutativity of multiplication then it must be that, for example, $2 \cdot (-3) = (-3) \cdot 2$. Extending other rules, we can also reduce all operations to operations with natural numbers. For example, let us calculate $2 \cdot (-3)$. We will show that $2 \cdot (-3)$ is the opposite number of $2 \cdot 3$:

$$2 \cdot 3 + 2 \cdot (-3) = (\text{distributivity}) = 2 \cdot (3 + (-3)) = 2 \cdot 0 = 0$$

So, $2 \cdot (-3) = -(2 \cdot 3) = -6$. Thus, we have obtained the same rule for multiplication that we obtained from previous content-wise considerations.

Content-wise and Structural in Mathematics

Sometimes we define new elements of a mathematical story and sometimes we just introduce them as new objects with intended properties in the structure of the mathematical story.

Basic Properties of Multiplication of Integers

1. $a \cdot b = b \cdot a$ (**commutativity**)

2. $(a \cdot b) \cdot c = a \cdot (b \cdot c)$ (**associativity**)

3. $1 \cdot a = a$ (**neutrality of one**)

4. $0 \cdot a = 0$

5. $a \cdot (b + c) = a \cdot b + a \cdot c$ (**distributivity of multiplication over addition**)

6. *For $a \neq 0$, if $a \cdot b = a \cdot c$ then $b = c$* (**cancellation by non-zero number**)

Multiplication, as we have defined it, fulfils these properties. Moreover, it may be shown that any differently defined multiplication would not fulfil these properties (structural approach!). Here is the proof of associativity, and you may see in a similar way (or prove) that the other properties hold as well. When we multiply three numbers, the amount of the result equals the product of the amounts of these three numbers, and it does not depend on the order of multiplication. Also, in determining the sign, an odd number of negative signs will yield a negative sign; otherwise, the sign will be positive. Thus, it does not depend on the order of multiplying.

Since the addition and multiplication of integers have the same basic properties as addition and multiplication of natural numbers, *any transformation of the description of numbers that we have performed with natural numbers holds now as well*. For instance,

$$2a + 4a = 6a, \; 2a \cdot 4a = 12a^2, \; 2 \cdot (a + 4) = 2a + 8, \text{ etc.}$$

(the definition of raising to the power as shortened multiplication remains the same).

However, what is to be done if somewhere there appears the sign '−', for example, $2a - 4a$? We have simplified the description $2a + 4a$ using the law of distributivity:

$$2a + 4a = a \cdot (2 + 4) = 6a$$

In the same way we can also solve $2a - 4a$, since the law of distributivity holds in such cases as well. For example,

$$a \cdot (b - c) = a \cdot (b + (-c)) = a \cdot b + a \cdot (-c) = a \cdot b - a \cdot c$$

Distributivity of Multiplication to Addition and Subtraction

$$a \cdot (b - c) = a \cdot b - a \cdot c$$
$$a \cdot (-b + c) = -a \cdot b + a \cdot c$$
$$a \cdot (-b - c) = -a \cdot b - a \cdot c$$

Thus, $2a - 4a = (2 - 4)a = -2a$.

Example 2.7.2. Let us simplify:

1. $2a - 4b - 6a + 6b$

2. $2x - (3y - 4x) + (-x + 2y)$

3. $-2(x - 3y) + 4(x - 2y)$

4. $(2a - 3b) \cdot (-3a + 2b)$

5. $(-a)^2 \cdot a^3 + a^2 \cdot (-a)^3$

Solution.

1. $2a - 4b - 6a + 6b = (2-6)a + (-4+6)b = -4a + 2b$

2. $2x - (3y - 4x) + (-x + y) = 2x - 3y + 4x - x + 2y = 5x - y$

3. $-2(x - 3y) + 4(x - 2y) = -2x + 6y + 4x - 8y = 2x - 2y$

4.

$$(2a - 3b) \cdot (-3a + 3b) = -6a^2 + 4ab + 9ab - 6b^2 = -6a^2 + 13ab - 6b^2$$

5. It is easy to see that the following holds:

Sign of the Power

$(-a)^n = -a^n$ if n is an odd natural number

$(-a)^n = a^n$ if n is an even natural number

Thus,

$$(-a)^2 \cdot a^3 + a^2 \cdot (-a)^3 = a^2 \cdot a^3 - a^2 \cdot a^3 = 0$$

$\square$

The relationship of comparison and multiplication is as follows:

Comparison and Multiplication

For $a > 0$, $ba < ca$ precisely when $b < c$

For $a < 0$, $ba < ca$ precisely when $b > c$

It is stated that multiplication and cancellation by a positive number preserves the comparison, and multiplication and cancellation by a negative number reverses the comparison.

Example 2.7.3. Let us prove: *for $a > 0$, if $b < c$ then also $ba < ca$.*

Solution. Let $b < c$. Then according to the compatibility of addition and comparison also $b + b < c + b$, but also $c + b < c + c$. It follows that $b + b < c + c$. By repeating a times this procedure, we obtain that

$$b + \ldots + b \text{ (} a \text{ members are added)} < c + \ldots + c \text{ (} a \text{ members are added),}$$

$$\rightarrow \quad ba < ca.$$

Similarly (by subtracting) we would prove also the opposite: that for $a > 0$, from $ba < ca$ there follows $b < c$. Only then would we prove the first rule of compatibility of multiplication and comparison.

The division of an integer by a natural number can be introduced in the same way as in the case of natural numbers. Just as $6 : 2 = 3$, so we can take that $(-6) : 3 = -2$. A debt of 6 kuna has been equally divided into 3 equal parts. However, what will $(-6) : (-3)$ be? In the same way as in multiplication, we will understand this as a combination of two actions: the change of the sign and dividing by 3. Thus, we will establish that it is $(-6) : (-3) = (-(-6)) : 3 = 6 : 3 = 2$.

Division of Integers

Dividing by a positive number is the same as dividing by a natural number, there is no dividing by zero, and dividing by a negative number means changing the sign and dividing by its amount. Division conceived in this way, when feasible, yields a result whose amount equals the quotient of the amounts of the numbers that are to be divided:

$$a : b = \pm |a| : |b|$$

What sign is to be used is determined in the same way as in the case of multiplying:

- *if a and b have the same signs, the result has the sign $+$;*

- *if a and b have different signs, the result has the sign $-$.*

It is easy to see that, defined in this way, division is an operation inverse to multiplication.

> **Inversion of Multiplication and Division**
>
> $a : b = c$ *precisely when* $c \cdot b = a$

Division can also be introduced in a structural way, as an operation inverse to multiplication.

Example 2.7.4. In his will, the late Signor Domnius divided the money from his current account. Since in *Our Little Town* male children are preferred, he left his son half of the money, and his two daughters a quarter each. However, as it happened, the balance of his account was $-80\,000$ kuna. How did the heirs divide the inheritance?

Solution. The son gets $-80\,000 : 2 = -40\,000$ kuna, and each of the daughters $-80\,000 : 4 = -20\,000$ kuna. Of course, the negative sign indicates that they have inherited their father's debts. $\qquad\square$

You can acquire additional knowledge on integers at `https://en.wikip edia.org/wiki/Negative_number`. Especially interesting is the "negative" history of negative numbers. The acceptance of negative numbers was a very slow process because there is not such a direct connection with "real" things as in the case of natural numbers.

2.8   Problems and Solutions

Comparison

1. Order the numbers $-32, 9, 34, 0, -17, 23$

2. Are the following statements true?

   (a) The smallest integer does not exist.

   (b) Any non-empty set of integers has the smallest element.

   (c) Between any two integers there is an integer.

Addition

3. If from position -7 you move by (-12) at what position will you arrive?

4. Before the basketball match, *Vertigo* BC had a score difference of -86. The club lost the match with a 30 point difference. How much is the score difference now?

5. When it saw Terrible Joe, the vulture flew agitatedly off the floor of Death Valley (altitude above sea level: -282 of cowboy feet) 500 cowboy feet upwards. At what altitude above sea level is it now?

6. The residents of the Ice Planet like the cold. The colder, the better! Therefore, they use positive numbers to mark states colder than $0\ {}^\circ\text{C}$, and they use negative numbers to mark warmer states. Is this going to affect their definition of adding integers? If their temperature was $-10\ {}^\circ\text{C}$, and then increased by $15\ {}^\circ\text{C}$, what is it now? Will their ice cream melt at that temperature?

Opposite Number

7. Are the following statements true?

   (a) $|-a| = -|a|$

   (b) $-(-a) = |a|$

   (c) $a + (-a) = 1$

8. Using the definition of addition and the concept of opposite number, prove that $a + (-a) = 0$.

Subtraction

9. Before the match, *Ball* FC had a goal difference of -7 goals, and after the match the goal difference was -11 goals. Did they win or lose the match, and with how many goals difference?

10. The water level of the Čikola (a river in Croatia) near Čavoglave (a village in Croatia with the unusual phenomenon that many people see things there in an exaggerated manner) has risen alarmingly from -73 cm to -62 cm. How much has the water level risen?

Addition and Subtraction

11. Find:

    (a) $3 - (-7) - (-9) + (-8)$ (b) $3 - (-(-2) + (-(-4)))$
    (c) $-(2 - 4) - (-2 - (2 - 3))$

12. If the basketball club *Vertigo* BC had a score difference of -56, and subsequently they lost one match with 14 points difference, and won another with 4 points difference, what is the score difference now?

13. The elevator in the *Better Life* mine started from the tenth level below ground and went 3 levels upwards, 5 downwards and then 6 upwards. At what level is it now?

14. It has been agreed that an electron has a negative charge of -1, and a proton has a positive charge of 1. At the beginning of the experiment there were -7 units of charge on the device. First 20 electrons and 15 protons were added, and then 10 electrons were removed. What is the present total charge in the device?

15. Prove: If $a < b$ then $-a > -b$.

Multiplication

16. Find:

    (a) $(-3)\cdot(-2)-4\cdot(-2)$ (b) $(-3-(-1)\cdot(-3))\cdot((-2)\cdot(-1)-2)$

17. Simplify and check the result using the software *SageMath*:

    (a) $-2x-(-2y-3x)+(-4x+2y)$ (b) $-2b(-3a+2ab)-3a(-2b^2-a)$
    (c) $(a-b)\cdot(-3a-2b)$ (d) $-(-a)^2\cdot(-a)^3-a^2\cdot(-a)^3$

18. Romeo's current account balance was -7300 kuna. He raised a loan from the bank of $20\,000$ kuna. For this service the bank charged a one-off fee of 5200 kuna. Romeo pays back to the bank each month 900 kuna. What will be the balance of his account after one year, if other inflows and outflows on the account are mutually cancelled?

19. At an altitude of 3000 cm a seagull saw a young anchovy at a depth of 10 cm. If the gull swoops down at a speed of 90 cm per second, and the anchovy swims deeper at a speed of 1 cm per second, at what mutual distance will they be in 30 seconds?

Various Problems

20. Solve in the simplest possible manner:

    (a) $-31+28-25+30$ (b) $23\cdot(-8)-23\cdot(-7)$

21. Simplify and check the result using the software *SageMath*:

    (a) $(5-a)\cdot(2+b)-(3-a)\cdot(2-b)$ (b) $-3ab-a\cdot[2-3\cdot(b-a)]$

22. The floating restaurant *Storm* leaks on all sides. It fills up with 5 litres of seawater per minute. If 10 waiters bail out the water using aperitif glasses of 1 decilitre, at a rate of 2 glasses per minute, how much seawater will be aboard in half an hour?

23. The balance of John's current account near the end of the month was −436 kuna. In a few days his income of 4456 kuna was deposited in the account, and then his son (who owns an additional card) withdrew at an ATM 500 kuna three times, and his daughter in her usual tour around the town drained her additional card of a mere 2347 kuna, and his wife, wanting to surprise him (also using an additional card linked to his account), paid for a seven-day cruise for a good price of 7445 kuna. "Fortunately, travel expenses of 230 kuna were deposited on the account!" thinks John. What is the current balance on John's account?

Solutions

1. $-32, -17, 0, 9, 23, 34$

2. (a) Yes   (b) No (e.g. $\{-2, -4, -6, \ldots\}$)   (c) No

3. $-7 + (-12) = -19$

4. $-86 + (-30) = -116$

5. $-282 + 500 = 218$ (cowboy fees)

6. The fact that the Ice Planet residents will choose for the positive orientation the opposite orientation than we do will not affect the definition of addition. It will affect only the interpretation of numbers. It will always be that $-10 + 15 = 5$. However, they find the temperature of $-10\,°\mathrm{C}$ the temperature at which ice cream will melt, and the adding of the positive number 15 means a change in the positive direction, which means in their case towards a colder condition. The result of addition is the positive number 5, and it measures a condition colder than water freezing, a temperature at which ice cream will not melt.

7. (a) No   (b) No, if a is a negative number, otherwise yes   (c) No

8. The numbers a and $-a$ have the same amounts ($|-a| = |a|$), but opposite signs. According to the rule of addition of numbers with opposite signs, their sum will be obtained by subtracting their amounts. Since their amounts are the same, the result is zero.

9. $-11 - (-7) = -4$. Thus, they lost the match with 4 goals difference.

10. $-62 - (-73) = 11$. Water level rose by 11 cm.

11. (a) 11   (b) -3 (c) 3

12. One method is to add to the initial goal difference the changes that can be positive or negative: goal difference $= -56 + (-14) + 4$. The second method is to add and subtract from the initial goal difference constantly positive amounts: goal difference $= -56 - 14 + 4$. Of course, in both cases the result is the same and it amounts to -66.

13. $-10 + 3 - 5 + 6 = -6$

14. $-7 + (-20) + 15 - (-10) = -2$

15. From $a < b$ according to compatibility of comparison and addition it follows that $a + (-a) < b + (-a)$, i.e. $0 < b + (-a)$. In the same way we further have $0 + (-b) < b + (-a) + (-b)$, i.e. $-b < -a$. Thus, $-a > -b$.

16. (a) 14   (b) 0

17. (a) $-3x + 4y$ (b) $6ab + 2ab^2 + 3a^2$ (c) $-3a^2 + ab + 2b^2$ (d) $2a^5$

18. $-7300 - 20\,000 - 5200 + 12 \cdot 900 = -21\,700$. He will still owe the bank 21700 kuna.

19. The position of the gull is $3000 - 90 \cdot 30 = 300$ cm, and the position of the anchovy is $-10 - 1 \cdot 30 = -40$ cm. Thus, they are at a mutual distance of $|-40 - 300| = |-340| = 340$ cm. Good luck, little fish!

20. (a) 2   (b) -23

21. (a) $4 + 8b - 2ab$ (b) $-2a - 3a^2$

22. The ship will contain 90 litres of seawater.

23. -7052 kuna

3

Rational Numbers
(or how to divide evenly)

In this chapter we will revise our knowledge about rational numbers:

- how we imagine them and what their purpose is;

- how we write them down and present them on a straight line;

- why and how we compare, add, subtract, multiply, divide and raise them to the power;

- what the properties of these operations with rational numbers are and why they are better than the properties of these operations with integers.

We will recall equations, which are very powerful mathematical tools for describing and resolving problem situations, as well as equalities, with which we express the laws of nature. We will see that, unlike natural and integer numbers, only rational numbers allow easy work with equations and equalities.

But let me warn you, in this chapter we will also hang out with pirates.

3.1 Measuring of Equal Parts

Three pirates, Giorgio, Giovanni and Francesco stole two equal golden bars and they want to divide them evenly. If they used natural numbers, that would be very bad for them: $2 \div 3 = 0$ and $2 \bmod 3 = 2$. Thus, everybody would get zero bars, and the two bars would remain undivided. In order to avoid new bloodshed, we will help them with the division. Since the bars are of a regular form, they need to be divided crosswise into three equal parts. The division can be performed geometrically in the following manner:

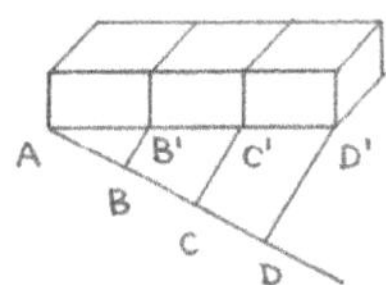

From point A we will draw a ray in an arbitrary direction and apply a certain length $\overline{AB}$ (e.g. with a compass) to it three times. We connect point D with D' and parallel with length $\overline{DD'}$ we draw the lengths $\overline{CC'}$ and $\overline{BB'}$. The geometry ensures that the golden bar is divided with points B' and C' into three equal parts.

We could have divided the bar equally into any (natural) number of parts. The division into six parts yields smaller parts than the division into three parts. Moreover, two such smaller parts are of equal value to one part obtained by dividing into three parts.

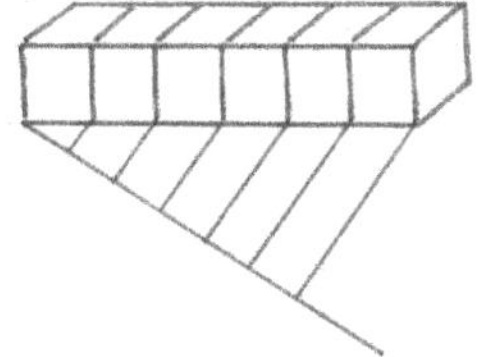

The comparison of parts, and addition and subtraction of parts, will be made easier to perform by the introduction of new numbers. We will use them to measure such parts (created by dividing the whole into equal parts). The part created by dividing the bar into three equal parts is completely determined by the number 3. The pirates will divide also the other bar in the same way, so that each pirate will get a total of two such parts. Thus, each share of the loot is completely defined by two natural numbers: the number 3, which says which part of the unit is taken, and the number 2, which says how many times it has been taken. It is precisely these two numbers that define the measure of the loot, a new number that is completely defined by them. Therefore, we will designate it with $\frac{2}{3}$ and read *two thirds*. In this way each pirate will get his share of the loot. The division is fair because every part has an equal measure, the number $\frac{2}{3}$:

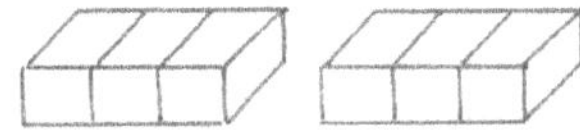

Naturally, the bar can be divided in a simpler manner with fewer cuts. Instead of the previously described 4 cuts, the loot can be divided also by two cuts (but I am not sure the pirates would like that):

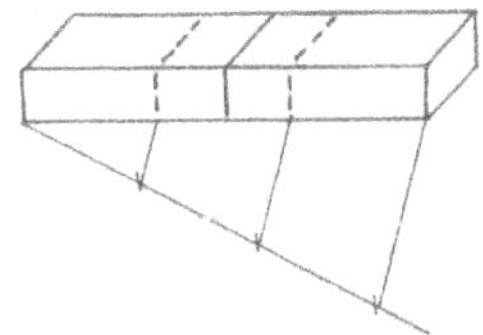

Such division shows us another property of the obtained number. The number 2 can be divided by the number 3 without remainder, and the result of this operation is a new type of number:

$$2 : 3 = \frac{2}{3}$$

And so the pirates – happy and satisfied – divide the loot. And what have we got? Knowledge. Knowledge is the gold bar in our head, if we know how to use it. And unlike the actual gold bar, nobody can steal it from us.

In the same way, we can divide m bars into n equal parts. The measure of every part is the new number $\frac{m}{n}$. This is the measure of the part that we will get when the n-th part of a unit bar is taken m times. This is at the same time the result of the division of natural numbers m and n:

$$m : n = \frac{m}{n}$$

The n-th part of zero bars gives zero bars. Thus, $\frac{0}{n} = 0$. If we imagine that the pirates have opened their own bank, *Pirate Bank* Ltd., in which the currency is not kuna, but rather gold bars, then the debts will be also measured in gold bars.

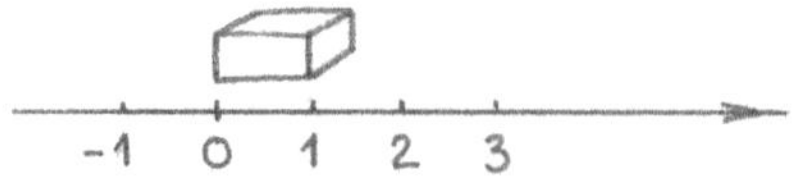

So, a pirate can even owe $\frac{2}{3}$ bars, that is to say, the balance of his account will be $-\frac{2}{3}$.

Rational Numbers

The measures of parts formed by division of the whole into equal parts will be called **rational numbers** and their set will be denoted by $\mathbb{Q}$. Any such rational number is described by one integer m and one positive integer n. The number n says which part of one should be taken, and m how many times this has to be done, maybe even in the negative direction, in order to obtain this rational number.

For any rational number a there is an integer m and a positive integer n such that

$$a = \frac{m}{n}$$

(the connection of rational numbers with integers)

This rational number is at the same time the result of the division of numbers m and n:

$$m : n = \frac{m}{n}$$

Such a notation of a rational number is called notation in the form of **fraction**. *The upper* number is called the **numerator**, and *the lower* the **denominator** of the fraction.

If we divide the integer m into one part, we will obtain again the number m. Thus, the number $\dfrac{m}{1}$ measures the same value as the number m. Therefore, the following rule is valid:

> ### Integers are Rational Numbers
>
> For any integer m, $m = \dfrac{m}{1}$

Rational numbers can be vividly presented by points on a straight line. If we choose on a line the zero point (0), positive direction (+) and a unit step (1):

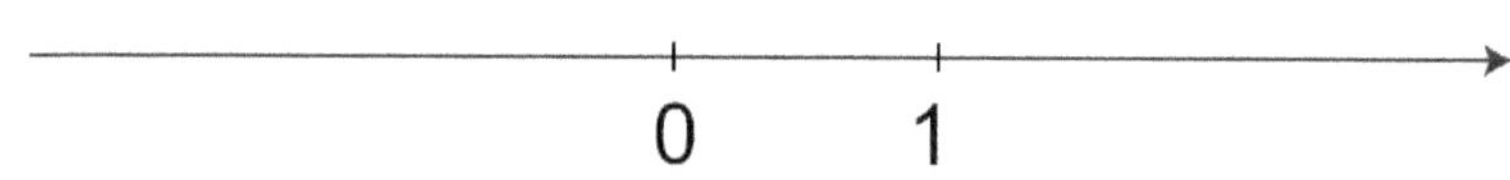

then, for instance, the rational number $\dfrac{4}{3}$ is assigned the point that is obtained when a third of the unit length is applied four times in the positive direction:

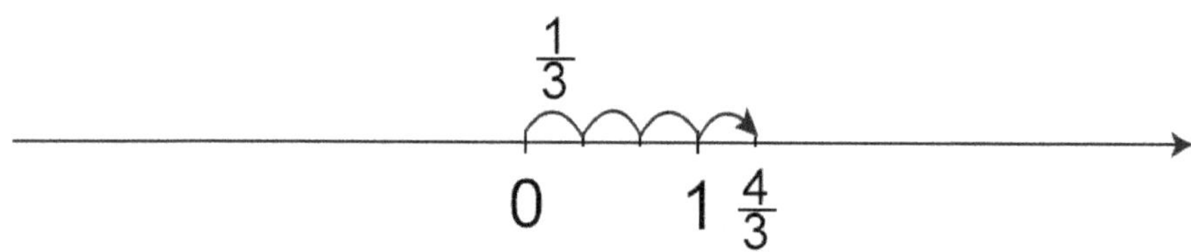

In the same way the rational number $\dfrac{-3}{2}$ is assigned the point obtained when a half of the unit length is applied in the negative direction three times:

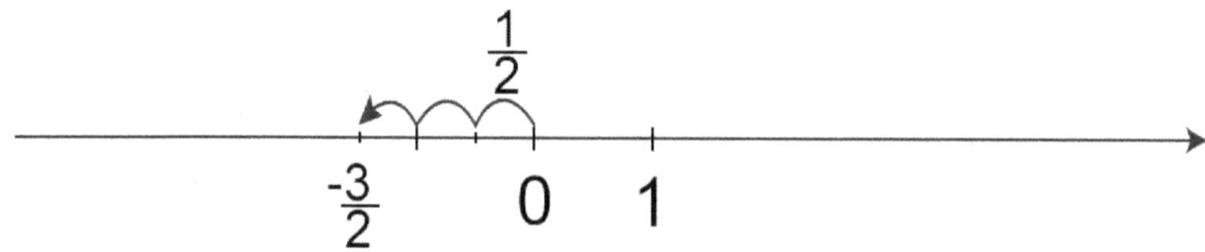

Graphical Representation of Rational Numbers

On a line with a selected origin, positive direction and unit length, the rational number $\dfrac{m}{n}$ is assigned a point which is obtained when by starting from the origin, we apply an n-th part of the unit length $|m|$ times in the positive direction for m positive, and in the negative direction for m negative number:

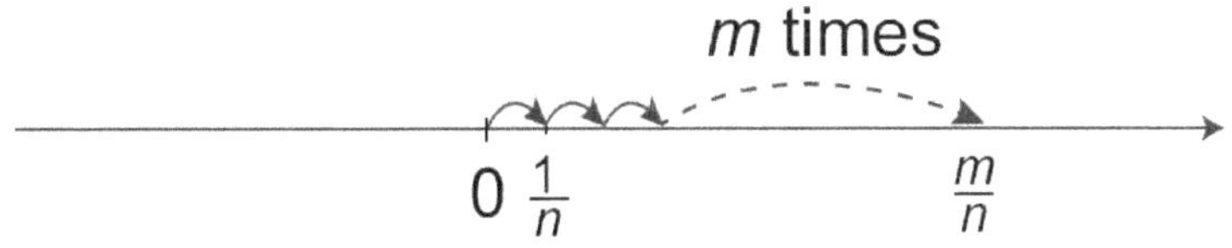

It has been already mentioned in the division of gold bars that the same rational number can be expressed by integers in several ways. Let us look at the rational number $\dfrac{m}{n}$. If a unit has been divided into k times more parts ($n \cdot k$), such a part needs to be applied k times more ($m \cdot k$) in order to obtain the same number.

Expanding and Reducing (Cancelling) Fractions

$$\frac{m}{n} = \frac{m \cdot k}{n \cdot k}, \text{ for } k \neq 0$$

If the numerator and the denominator of the fraction are multiplied by the same non-zero natural number, or divided by a non-zero natural number with which the numerator and denominator are both divisible, we will obtain the same rational number.

For instance, $\dfrac{9}{6} = \dfrac{9 \cdot 2}{6 \cdot 2} = \dfrac{18}{12}$, but in the same way also $\dfrac{9}{6} = \dfrac{9 : 3}{6 : 3} = \dfrac{3}{2}$.

The notation $\dfrac{3}{2}$ seems better than the notation $\dfrac{18}{12}$, since it consists of smaller numbers. It cannot be any more *reduced* because 3 and 2 have no common divisor greater than 1. Such a notation, in which the numerator and the denominator have no common divisor greater than 1, is called an **irreducible fraction**. Other fractions are called **reducible fractions**. Naturally, from a reducible fraction we will get an irreducible fraction by dividing (we say also **reducing** or **cancelling**) the numerator and denominator by a non-trivial common divisor, as long as it exists. This can be also done in one step, so that we divide them immediately by the biggest common divisor.

Example 3.1.1. Let us calculate:

$$1.\ \frac{240}{320} \qquad 2.\ \frac{108}{72} \qquad 3.\ \frac{2^5 \cdot 3^2 \cdot 5 \cdot 7^7}{2^3 \cdot 3^4 \cdot 7^7}$$

Solution.

1. We are searching for common divisors:

$$\frac{240}{320} = \frac{240:10}{320:10} = \frac{24}{32} = \frac{24:2}{32:2} = \frac{12}{16} = \frac{12:4}{16:4} = \frac{3}{4}$$

2. Instead of using the common divisors to divide gradually, we can find the biggest common divisor and divide at once using it. For instance, the biggest common divisor of numbers 72 and 108 is 36, so that

$$\frac{108}{72} = \frac{108:36}{72:36} = \frac{3}{2}$$

3. One need not calculate the numerator and the denominator and then search for common divisors, as they can be found from the given products. These are common factors of the numerator and denominator. In such notation the rule for reducing the fraction is:

   If the factor of the numerator and the factor of the denominator are divided by the common divisor, we will obtain the same rational number.

   Since the common divisor of the numbers 2^5 and 2^3 is the smaller one, that is 2^3, we will delete (reduce to 1) the smaller one, and diminish the exponent of the bigger one by the exponent of the smaller one. Thus,

$$\frac{2^{5^2} \cdot 3^2 \cdot 5 \cdot 7^7}{2^3 \cdot 3^{4^2} \cdot 7^7} = \frac{2^2 \cdot 5}{3^2} = \frac{20}{9}$$

$\square$

Example 3.1.2. Let us simplify:

$$1.\ \frac{m^3 n^2 k^4}{m^2 n^3 k} \qquad 2.\ \frac{km + kn}{mnk} \qquad 3.\ \frac{6m^3 n^3 k + 4m^2 n^4}{2m^3 n^3}$$

Solution.

1. We will reduce the common factors:

$$\frac{m^3\,n^2k^{\prime 3}}{m^2n^3\,k} = \frac{mk^3}{n}$$

2. In order to be able to cancel, first we have to factorise the numerator (present it as a product). Here we have the help of the distributive law $k(m+n) = km + kn$. In one direction ($\rightarrow$) it says how to get rid of the brackets. In the other direction ($\leftarrow$) it tells us that the description $km + kn$ can be factorised by taking out of the brackets the common factor k from both products:

$$\frac{km+kn}{mnk} = \frac{k(m+n)}{mn\,k} = \frac{m+n}{m\cdot n}$$

3. Again, we will factorise the numerator by taking out the common factor, trying to take out the greatest possible common factor. For the numbers 6 and 4, we take out the greatest common divisor, the number 2. In the numbers m^3 and m^2 we will take out the smaller power, m^2, and in the same manner from n^3 and n^4 we will take out the smaller power, n^3. The factor k is not a common factor, and therefore it cannot be taken out:

$$\frac{6m^3n^3k + 4m^2n^4}{2m^3n^3} = \frac{2\,m^2\,n^3(3mk + 2n)}{2m^3\,n^3} = \frac{3mk + 2n}{m}$$

We can try to simplify open descriptions with fraction in the software *SageMath*, with the command *simplify_full*. For instance, the last example can be solved in the following manner:

```
var('m,n,k')
show(((6*m^3*n^3*k+4*m^2*n^4)/(2*m^3*n^3)).simplify_full())
```

We will obtain

$$\frac{3km + 2n}{m}$$

3.2 Comparison

The existence of several notations for the same number raises the question of how one is to know on the basis of notation whether two numbers are the same or not. One way is to see whether their irreducible fractions are equal. The other way is expansion to the common denominator – the fractions are expanded so that they have the same denominators. Let us look, for instance, at the numbers $\frac{5}{12}$ and $\frac{3}{8}$. Our goal is to obtain the same denominator by expanding the fraction. It has to be a common multiple of the numbers 12 and 8. One common multiple is easy to determine – this is the product of the numbers $12 \cdot 8 = 96$. However, it is better to take a smaller common multiple so as not to get stuck with large numbers. Therefore, we will take the smallest common multiple. Although in the chapter on natural numbers we are reminded of the systematic procedures for finding the least common multiple, for smaller numbers it can be more easily determined by inspection. It is easy to determine that for numbers 12 and 8 this is precisely number 24. We will rewrite both fractions to have the same denominator by an appropriate expansion:

$$\frac{5}{12} = \frac{5 \cdot 2}{12 \cdot 2} = \frac{10}{24} \qquad \frac{3}{8} = \frac{3 \cdot 3}{8 \cdot 3} = \frac{9}{24}$$

Not only can we find out that these numbers are not equal, but we can also determine which one is larger. When we have, let us say, equal gold bars and their parts (or any other value consisting of equal wholes and their parts), using direct comparison it may be determined where there is more gold. We will transfer this comparison to their measures as well, so that the lesser rational number will mean less gold. If we speak about two fractions with the same denominators, the one having a greater numerator measures a greater quantity. Thus we obtain a simple definition of comparison.

Comparison of Rational Numbers

- $\dfrac{m_1}{n} = \dfrac{m_2}{n}$ *precisely when* $m_1 = m_2$
  (equality of rational numbers)

- $\dfrac{m_1}{n} < \dfrac{m_2}{n}$ *precisely when* $m_1 < m_2$
  (comparison of rational numbers)

This description enables a simple formal procedure of comparison.

Procedure of Comparison of Rational Numbers

Fractions are compared in the following way:

1. we expand them so that they have the same denominator;

2. we compare the numerators (this is the already known comparison of integers): the fraction with the greater numerator is greater.

On the basis of this description it is easy to prove that the comparison of rational numbers is the extension of the comparison of integers and that it has the following properties:

Basic Properties of Comparison of Rational Numbers

1. *For any number a there are numbers l and g such that $l < a < g$* (**existence of a lesser and a greater number**)

2. *If $a < b$ and $b < c$, then $a < c$* (**transitivity**)

3. *Either $a < b$ or $b < a$ or $a = b$* (**strict comparability**)

4. *For any two numbers a and b there is number c which is between them: if $a < b$, then $a < c < b$* (**density**)

Let us note here that, unlike integers, it is not stated that there exist the first greater and the first lesser rational number than a, since it is precisely the density property which says that there are no such numbers. Whichever number g we take which is greater than a, it cannot be also the first greater number than a since, regarding the property of density, there is the number c which is lesser than this number, and greater than a. Analogously, it is proved that there is no first number lesser than a.

Let us prove the property of density. Let $a < b$, $a = \dfrac{m_1}{n}$ and $b = \dfrac{m_2}{n}$. Then $m_1 < m_2$. If m_2 is not the first number after m_1 ($m_2 \neq m_1 + 1$), then $m_1 < m_1 + 1 < m_2$, so that $\dfrac{m_1}{n} < \dfrac{m_1 + 1}{n} < \dfrac{m_2}{n}$. Thus, $\dfrac{m_1 + 1}{n}$ is the required rational number which is between a and b. If, however, $m_2 = m_1 + 1$, we will double the denominators (look at the smaller parts):

$$\frac{m_1}{n} = \frac{2m_1}{2n}, \quad \frac{m_2}{n} = \frac{m_1+1}{n} = \frac{2(m_1+1)}{2n} = \frac{2m_1+2}{2n}$$

However, now $2m_1 < 2m_1 + 1 < 2m_1 + 2$, so that $\dfrac{2m_1}{2n} < \dfrac{2m_1+1}{2n} < \dfrac{2m_1+2}{2n}$

Thus, now $\dfrac{2m_1+1}{2n}$ is between a and b.

Example 3.2.1.

1. Let us compare the rational numbers $\dfrac{-3}{2}, 2, \dfrac{-5}{6}, \dfrac{7}{9}$.

2. Let us find a rational number a greater than $\dfrac{1}{7}$ and a rational number b greater than $\dfrac{1}{7}$, but lesser than a.

Solution.

1. The smallest common multiple of the denominators 2, 1, 6 and 9 is 18. We will write every fraction with this denominator:

$$\frac{-3}{2} = \frac{-27}{18}, \quad 2 = \frac{2}{1} = \frac{36}{18}, \quad \frac{-5}{6} = \frac{-15}{18}, \quad \frac{7}{9} = \frac{14}{18}$$

Comparing the numerators, we get $\dfrac{-27}{18} < \dfrac{-15}{18} < \dfrac{14}{18} < \dfrac{36}{18}$.

Thus, $\dfrac{-3}{2} < \dfrac{-5}{6} < \dfrac{7}{9} < 2$.

Larger than $\dfrac{1}{7}$ may be, for instance, the number $a = \dfrac{2}{7}$. However, whichever number a we take, regarding the property of density of rational numbers, there is a lesser number that is still greater than $\dfrac{1}{7}$. If it seems to us that between the numbers $\dfrac{1}{7}$ and $\dfrac{2}{7}$ there is "a little space", we can express that with tinier parts: $\dfrac{1}{7} = \dfrac{2}{14}, \dfrac{2}{7} = \dfrac{4}{14}$.

Thus, in between there is $\dfrac{3}{14}$. Naturally, we can take even smaller bits: $\dfrac{1}{7} = \dfrac{2}{14} = \dfrac{4}{28}, \dfrac{3}{14} = \dfrac{6}{28}$. Which means the number $\dfrac{5}{28}$ is even closer to $\dfrac{1}{7}$. Simply, *there is no first greater number after the given rational number*. Whichever number you propose, with this procedure I can find a lesser number which is greater than the given rational number.
                                                                                                              □

3.3　Addition and Subtracti

We described the addition and subtraction of natural numbers by adding and subtracting the quantities they measure. Equally, we can describe both the addition and subtraction of the non-negative rational numbers a and b:

$a + b$ is the measure of the quantity we obtain when we add the quantity of measure a to the quantity of measure b.

For $a > b$, $a - b$ is the measure of quantity that we obtain when we subtract the quantity of measure b from the quantity of measure a.

These operations can also be transferred to negative rational numbers, just as we transferred these operations with natural numbers to integers. Here, addition and subtraction numerically express addition and subtraction of changes of a quantity. Regardless of whether this refers to *Our Bank* or to *Pirate's Bank* Ltd., the procedure is the same: *addition is done by adding units and their parts to the current account, and subtraction is done by withdrawing them from the current account.*

On the basis of this description it is easy to find out how fractions with the same denominators are added and subtracted:

Addition and Subtraction of Rational Numbers

- $$\frac{m_1}{n} + \frac{m_2}{n} = \frac{m_1 + m_2}{n}$$ **(addition of rational numbers)**

- $$\frac{m_1}{n} - \frac{m_2}{n} = \frac{m_1 - m_2}{n}$$ **(subtraction of rational numbers)**

With these formulas, the addition and subtraction of rational numbers have been described using the already known addition and subtraction of integers. Therefore, we can take them for definitions of these operations. The formal procedure of calculation follows from them.

Procedure of Addition and Subtraction

We add (subtract) fractions by first expanding them to a common denominator, and then adding (subtracting) the numerators.

Example 3.3.1.

1. Let us calculate $\dfrac{6}{8} + \dfrac{10}{60}$.

2. In their last maritime battle, heroically defending their treasure stolen with great effort, both pirates Stefano and Matteo were killed. In their will they left their share of the total robbed treasure to their co-fighter pirate Francesco. Since Stefano was a shareholder of the pirate society *Using Hook, Not Hoe,* $\dfrac{1}{12}$ of the robbed treasure belonged to him, and $\dfrac{1}{64}$ of the stolen treasure belonged to Matteo, as a regular front-line pirate. What share of the stolen treasure did Francesco inherit?

Solution.

1.

Reduce Fractions

It is useful to reduce fractions before calculation.

$$\frac{6}{8} + \frac{10}{60} = \frac{3}{4} + \frac{1}{6} = \frac{9}{12} + \frac{2}{12} = \frac{11}{12}$$

2. Francesco inherited $\dfrac{1}{12} + \dfrac{1}{64} = \dfrac{16}{192} + \dfrac{3}{192} = \dfrac{19}{192}$ of treasure. $\qquad\square$

The addition and subtraction of rational numbers extend the addition and subtraction of integers, keeping all the good properties of these operations.

> ### The Basic Properties of Addition and Subtraction of Rational Numbers
>
> 1. $a + b = b + a$ (**commutativity of addition**)
>
> 2. $(a + b) + c = a + (b + c)$ (**associativity of addition**)
>
> 3. $a + 0 = a$ (**neutrality of zero**)
>
> 4. $a - b = c$ *precisely when* $c + b = a$
>   (**inversion of addition and subtraction**)
>
> 5. $a + c < b + c$ precisely when $a < b$
>   (**compatibility of comparison and addition**)

All these rules can be easily proved from the definition of addition and subtraction and the properties of integers. We will prove, for instance, the property of commutativity.

Let $a = \dfrac{m_1}{n}$ and $b = \dfrac{m_2}{n}$. Then $a + b = \dfrac{m_1}{n} + \dfrac{m_2}{n} =$ (definition of addition) $\dfrac{m_1 + m_2}{n} =$ (commutativity of addition of integers) $\dfrac{m_2 + m_1}{n} =$ (definition of addition) $\dfrac{m_2}{n} + \dfrac{m_1}{n} = b + a.$

Since the basic rules of addition and subtraction are the same both for rational numbers and for integers, all other rules that we have derived and used in the case of integers are the same as well, for instance, how to calculate several additions and subtractions, how to get rid of brackets preceded by the sign '$-$' and so on. However, here is one more useful rule for the sign, and it says that it does not matter whether it is the sign of the numerator or the sign of the entire fraction.

> ### Sign of the Fraction
>
> $$\frac{-m}{n} = -\frac{m}{n}$$

In order to prove that, we need to show that $\dfrac{-m}{n}$ is the opposite number of number $\dfrac{m}{n}$, i.e. their sum yields zero. Let us calculate:

$$\frac{m}{n} + \frac{-m}{n} = \frac{m + (-m)}{n} = \frac{0}{n} = 0.$$

Example 3.3.2.

1. Let us calculate $-\dfrac{2}{3} + \dfrac{3}{4} - \dfrac{1}{6}$

2. Let us simplify:

(a) $\dfrac{3}{2n} - \dfrac{m-n}{6mn} + \dfrac{2n+1}{3n^2}$ (b) $\dfrac{1}{n+m} + \dfrac{1}{n^2+mn} - \dfrac{1}{n^2}$

Solution.

1. We could calculate the operations one by one, but it is simpler immediately to set everything to the common denominator 12:

$$-\frac{2}{3} + \frac{3}{4} - \frac{1}{6} = \frac{-8+9-2}{12} = -\frac{1}{12}$$

2. (a) We need to expand the fractions so that they have the same denominators. Out of $2n$, $6mn$ and $3n^2$ we need to make the smallest product which can contain all of them by multiplication. The procedure is the same as in searching for the smallest common multiple of the numbers:

   From any factor in the denominators we take the biggest exponent with which it appears and from these powers we make the product.

   In this case, along with the concrete numbers from which we take the smallest common multiple (6), the factors are also n and m. The biggest power in which n occurs is n^2, and in which m occurs is precisely m. Therefore, the requested common denominator is $6mn^2$. The numerators are expanded by what the denominator lacks to a common denominator:

$$\frac{3}{2n} - \frac{m-n}{6mn} + \frac{2n+1}{3n^2} = \frac{3 \cdot 3mn - n(m-n) + 2m(2n+1)}{6mn^2} =$$

$$\frac{9mn - nm + n^2 + 4mn + 2m}{6mn^2} = \frac{12mn + n^2 + 2m}{6mn^2}$$

(b) First, we will factorise the denominators, in order to form the smallest common denominator from the factors:

$$\frac{1}{n+m} + \frac{1}{n^2+mn} - \frac{1}{n^2} = \frac{1}{n+m} + \frac{1}{n(n+m)} - \frac{1}{n^2} =$$

$$\frac{n^2 + n - (n+m)}{n^2(n+m)} = \frac{n^2 + n - n - m}{n^2(n+m)} = \frac{n^2 - m}{n^2(n+m)}$$

3.4 Mixed Notation

If a pirate tells you that he has $\dfrac{119}{9}$ gold bars, it is difficult to understand from this notation how many there actually are. Therefore, we will write down this rational number so as to see how many whole parts it contains. There is certainly 1 bar here since $1 = \dfrac{9}{9}$, but two bars as well because $2 = \dfrac{18}{9}$ etc. We can see that the number of whole bars is precisely the biggest number which multiplied by 9 does not yield more than 119. Thus, this is $119 \div 9 = 13$. Now we can write down the rational number in the following manner:

$$\frac{119}{9} = \frac{13 \cdot 9 + 2}{9} = \frac{13 \cdot 9}{9} + \frac{2}{9} = 13 + \frac{2}{9}$$

In brief, we write $13\dfrac{2}{9}$ (we read *thirteen and two ninths*) and this notation is called mixed notation of the rational number.

Mixed Notation for Rational Numbers

This is the description of a positive rational number using the biggest natural number which is smaller than or equal to it (the so-called **whole part of the rational number**) and the "remainder": the difference between the number and its whole part (the so-called **proper part of the rational number**).

For example, the mixed notation $13\dfrac{2}{9}$ of the rational number $\dfrac{119}{9}$ means that its proper part is $\dfrac{2}{9}$.

Calculation of the Mixed Notation

The whole part of the positive rational number $\dfrac{m}{n}$ is the result of division with the remainder of numbers m and n.

The proper part is $\dfrac{r}{n}$, where r is the remainder in division.

$$\frac{119}{9} = \underbrace{13}_{\text{WHOLE PART}} \underbrace{\frac{2}{9}}_{\text{PROPER PART}}$$

(where $119 \div 9$ gives the whole part 13, and $119 \bmod 9$ gives the numerator 2 of the proper part)

Example 3.4.1.

1. Write down the rational number $\dfrac{23}{3}$ in mixed notation.

2. Write down $12\dfrac{3}{5}$ in the form of a fraction.

Solution.

1. Division with remainder of the numbers 23 and 3 yields the result 7 and remainder 2. Thus,

$$\frac{23}{3} = 7\frac{2}{3}$$

2. We obtain the fraction notation by simple addition of the the whole and proper part of the mixed notation:

$$12\frac{3}{5} = 12 + \frac{3}{5} = \frac{12 \cdot 5 + 3}{5} = \frac{63}{5}$$

$\square$

The mixed notation shows the value of a rational number better, but the notation in the form of a fraction is more suitable for calculation.

3.5 Multiplication

Multiplication of a rational number by a natural number is actually abbreviated addition:

$$\frac{5}{2} \cdot 3 = \frac{5}{2} + \frac{5}{2} + \frac{5}{2} = \frac{5 \cdot 3}{2} = \frac{15}{2}$$

We can also easily discover a fast rule for multiplication by a natural number – the denominator is copied, and the numerator is multiplied by the natural number. But what is multiplication by for instance $\frac{1}{2}$? We think of it as taking a half of the number: multiplying by a unit gives that number, so multiplying by half a smaller number should give a half smaller result. Thus:

$$\frac{5}{2} \cdot \frac{1}{2} = \text{half of } \frac{5}{2} = \text{the number which fits two times into } \frac{5}{2}.$$

This is obviously the number with the same numerator, but a double denominator:

$$\frac{5}{2} \cdot \frac{1}{2} = \frac{5}{2 \cdot 2} = \frac{5}{4}$$

Thus, multiplication with $\frac{1}{n}$ is taking the n-th part of the number. The result has the same numerator, but n times bigger denominator: $\frac{k}{l} \cdot \frac{1}{n} = \frac{k}{l \cdot n}$. By combining the previous two illustrative examples of multiplication we get the general rule.

Multiplication of Rational Numbers

Multiplication of a rational number $\frac{k}{l}$ by an arbitrary rational number $\frac{m}{n}$ is taking the n-th part of the number m times. The result is given by the following formula:

$$\frac{k}{l} \cdot \frac{m}{n} = \frac{k \cdot m}{l \cdot n} \quad \textbf{(multiplication of rational numbers)}$$

One can see from the formula that before multiplication we can cancel with the common divisor any numerator with any denominator, which often facilitates the job.

Example 3.5.1.

1. Let us calculate $\dfrac{-8}{10} \cdot \dfrac{9}{6}$.

2. According to the tax law on inherited treasure, pirate Giorgio has to give the pirate state $\dfrac{1}{7}$ of the inherited $\dfrac{19}{192}$ of the kidnapped treasure. How much treasure will he keep?

3. Pirates X, Y and Z (I dare not give their real names so they will not sue me for mental distress) combined their treasure into a new business project. X invested 7, Y invested 8, and Z invested 5 gold bars. Owing exclusively to their extremely good business competencies and skill in waving their sabres, they realised a total income of 3000 gold bars. However, they did not know how to divide it equitably, and during their quarrel their ship sank with the entire treasure. Even today they accuse each other in the columns of the weekly magazine *Modern Entrepreneur* of incorrect business behaviour. What would have been a just distribution of the income?

Solution.

1. We can reduce every fraction:
$$\frac{-8}{10}\cdot\frac{9}{6}=-\frac{4}{5}\cdot\frac{3}{2}=$$
(we can also cancel the first numerator with the second denominator)
$$=-\frac{2}{5}\cdot\frac{3}{1}=-\frac{6}{5}$$

2. Since the state will take $\frac{1}{7}$ of the inheritance, the pirate Giorgio will keep $\frac{6}{7}$ of the heritage, which means $\frac{19}{192}\cdot\frac{6}{7}=\frac{19}{32}\cdot\frac{1}{7}=\frac{19}{224}$ of the kidnapped treasure.

3. The share in income should have been equal to the share in investment. Since out of the invested $7 + 8 + 5 = 20$ golden bars, 7 belonged to pirate X, he would have $\frac{7}{20}$ of the investment. Therefore, he should have had also $\frac{7}{20}$ of the returned, thus $3000\cdot\frac{7}{20}=1050$ bars. In the same way we can calculate that pirate Y would have had $3000\cdot\frac{8}{20}=1200$ bars, and pirate Z would have had $3000\cdot\frac{5}{20}=750$ bars. $\square$

Example 3.5.2.

1. Let us calculate $\dfrac{3-2}{2+4}\cdot\dfrac{3\cdot2}{2\cdot4}$

2. Let us simplify $\dfrac{5m-2m}{5n-2n}\cdot\dfrac{nk}{mk+nk}$

Solution.

1. We may cancel only factors of the numerators and denominators:
$$\frac{3-2}{2+4}\cdot\frac{3\cdot2}{2\cdot4}=\frac{1}{6_2}\cdot\frac{3}{4}=\frac{1}{8}$$

2. First we try to factorise the numerators and denominators:
$$\frac{5m-2m}{5n-2n}\cdot\frac{nk}{mk+nk}=\frac{3m}{3n}\cdot\frac{nk}{k(m+n)}=\frac{m}{m+n}$$

3.6 Division

Rational numbers enable division of an integer by a positive integer: $m : n = \dfrac{m}{n}$. This is the measure of the n-th part of the number m. For instance, $\dfrac{18}{5}$ is a fifth part of the number 18. However, this is also the number which fits 5 times into 18. In the same way we could also describe the division of a rational number by a natural number. For instance $\dfrac{18}{5} : 2$ is half of this number. In the previous topic we saw that multiplication of a number by $\dfrac{1}{2}$ is taking a half of the number. Thus $\dfrac{18}{5} : 2 = \dfrac{18}{5} \cdot \dfrac{1}{2}$. But what, for example, is $\dfrac{18}{5} : \dfrac{2}{3}$? It makes no sense any more to speak about the division of the number into $\dfrac{2}{3}$ equal parts, but we can continue to speak about the number which fits $\dfrac{2}{3}$ times into $\dfrac{18}{5}$ (its third part fits 2 times). Thus we think of division as an operation inverse to multiplication. How can we find $\dfrac{18}{5} : \dfrac{2}{3}$? We have seen that $\dfrac{18}{5} : 2 = \dfrac{18}{5} \cdot \dfrac{1}{2}$. Now we divide by a 3 times smaller number, so that division has to yield a 3 times bigger number (so as to fit 3 times less into $\dfrac{18}{5}$). Thus:

$$\frac{18}{5} : \frac{2}{3} = \frac{18}{5} \cdot \frac{3}{2}$$

Division of Rational Numbers

Division of rational numbers is an operation inverse to multiplication:

$\dfrac{k}{l} : \dfrac{m}{n}$ is a number which fits $\dfrac{m}{n}$ times into $\dfrac{k}{l}$.

For $\dfrac{m}{n} \neq 0$ this number exists and it is equal to $\dfrac{k}{l} : \dfrac{m}{n} = \dfrac{k}{l} \cdot \dfrac{n}{m}$.

We say for $\dfrac{n}{m}$ that it is the **reciprocal number** (lat. *reciprocus - inverse*) of number $\dfrac{m}{n}$. For negative m we will define that $\dfrac{n}{-|m|} = \dfrac{-n}{|m|}$.

Thus, *division by a non-zero number is equal to multiplication by its reciprocal number.*

Example 3.6.1.

1. Let us calculate: (a) $\dfrac{4}{5} : 2$, (b) $\dfrac{4}{5} : \dfrac{-2}{15}$

2. During their last business venture, 1658 litres of water entered into the pirates' ship, and for bailing it out the pirates have only a $\dfrac{9}{5}$ litre bucket which they bought on discount at *Pevex* (a Croatian retail chain). How many times will they have to use it to scoop up water in order to throw all the water out of their ship?

3. Let us simplify $\dfrac{m^2 n^3}{k l^2} : \dfrac{mn}{kl}$.

4. If we connect two resistors with resistances R_1 and R_2 in parallel, as in the figure below,

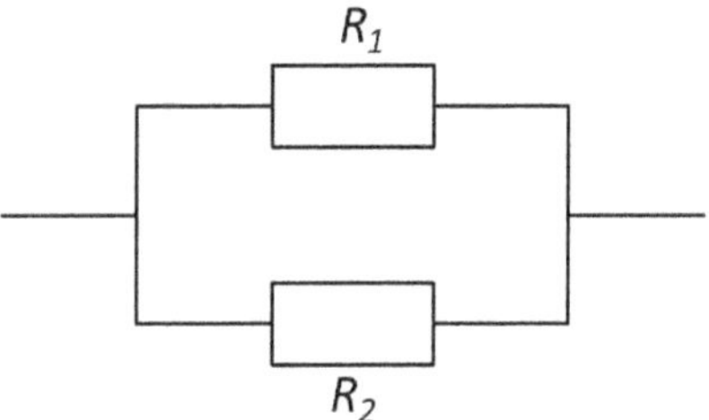

the total resistance of the connection R is given by the following formula: $\dfrac{1}{R} = \dfrac{1}{R_1} + \dfrac{1}{R_2}$. Let us use this formula in order to calculate the total resistance R if we connect in parallel two resistors with resistances $R_1 = 40$ Ohms and $R_2 = 10$ Ohms.

Solution.

1. (a) The reciprocal number to 2 $\left(= \dfrac{2}{1}\right)$ is $\dfrac{1}{2}$, so that

$$\frac{4}{5} : 2 = \frac{\cancel{4}^{2}}{5} \cdot \frac{1}{\cancel{2}} = \frac{2}{5}$$

(b) Since division is equal to multiplication by reciprocal number, the rules for signs are the same as for multiplication: an odd number of negative factors yields a negative result, and an even number yields a positive result. Thus we can immediately solve the issue of the sign:

$$\frac{4}{5} : \frac{-2}{15} = -\frac{\cancel{4}^2}{\cancel{5}} \cdot \frac{\cancel{15}^3}{\cancel{2}} = -6$$

2. We will calculate how many times $\dfrac{9}{5}$ litres fit into 1658 litres of water:

$$1658 : \frac{9}{5} = 1658 \cdot \frac{5}{9} = \frac{8290}{9} = 921\frac{1}{9} \text{ times.}$$

Thus, the pirates will have to fill the bucket 922 times (true, the last time only 1/9 of the bucket).

3. By transforming division into multiplication by a reciprocal number, we will simplify according to the rules for multiplication:

$$\frac{m^2 n^3}{kl^2} : \frac{mn}{kl} = \frac{m^2 n^3{}^2}{kl^2} \cdot \frac{\cancel{k}\,\cancel{l}}{\cancel{m}\,\cancel{n}} = \frac{mn^2}{l}$$

4. Let us include the known values of resistance in the formula and let us calculate:

$$\frac{1}{R} = \frac{1}{40} + \frac{1}{10} = \frac{1+4}{40} = \frac{5}{40} = \frac{1}{8}$$

The total resistance is a reciprocal number, $R = 8$ Ohm.   □

3.7 Multiplication and Division

By introducing the rational numbers we obtained an even better mathematical story than with integers. *We succeeded in expanding the multiplication of integers onto multiplication of rational numbers so that it keeps all the good properties of multiplication of integers and it is always possible to perform division by a non-zero number.*

Basic Properties of Multiplication of Rational Numbers

1. $a \cdot b = b \cdot a$ (**commutativity**)

2. $(a \cdot b) \cdot c = a \cdot (b \cdot c)$ (**associativity**)

3. $1 \cdot a = a$ (**neutrality of one**)

4. $0 \cdot a = 0$

5. $a \cdot (b + c) = a \cdot b + a \cdot c$
   (**distributivity of multiplication over addition**)

6. $a : b = c$ *precisely when* $c \cdot b = a$
   (**inversion of multiplication and division**)

7. *For* $c > 0$, $a \cdot c < b \cdot c$ *precisely when* $a < b$

   For $c < 0$, $a \cdot c > b \cdot c$ *precisely when* $a < b$
   (**compatibility of multiplication and comparison**)

These properties are easy to prove from the definition of multiplication and division of rational numbers and appropriate properties of integers.

If we compare these basic properties with the basic properties of multiplication of integers, we see that the cancellation law has been left out:

$$\text{For } a \neq 0, \text{ if } b \cdot a = c \cdot a, \text{ then also } b = c.$$

It has not been left out because it is not valid now, but because it easily follows from the inversion of multiplication and division.

Let $a \neq 0$ and $b \cdot a = c \cdot a$. Then

$$(b \cdot a) : a = (c \cdot a) : a \quad \text{e.g.} \quad b \cdot (a \cdot \frac{1}{a}) = c \cdot (a \cdot \frac{1}{a}), \text{ by associativity.}$$

It is easy to deduce from the sixth law that $a \cdot \dfrac{1}{a} = 1$. Thus, $b \cdot 1 = c \cdot 1$ and, by the neutrality of one, $b = c$.

It follows that in calculation with rational numbers we have the same freedom as in calculation with integers, and as well as that, the additional freedom which follows from the reduction of division into multiplication by reciprocal number.

Example 3.7.1.

1. Let us calculate $\dfrac{3}{5} \cdot \dfrac{7}{8} - \dfrac{2}{5} \cdot \dfrac{7}{8}$

2. Let us simplify $\dfrac{m}{n} : \dfrac{n}{k} : \dfrac{k}{n} \cdot \dfrac{n}{m}$

Solution.

1. We apply the law of distributivity:

$$\frac{3}{5} \cdot \frac{7}{8} - \frac{2}{5} \cdot \frac{7}{8} = \frac{7}{8} \cdot \left(\frac{3}{5} - \frac{2}{5} \right) = \frac{7}{8} \cdot \frac{1}{5} = \frac{7}{40}$$

2. The order is not important, because we consider division as multiplication by reciprocal number:

$$\frac{m}{n} : \frac{n}{k} : \frac{k}{n} \cdot \frac{n}{m} = \frac{m}{n} \cdot \frac{k}{n} \cdot \frac{n}{k} \cdot \frac{n}{m} = 1$$

$\square$

The integer m and the positive integer n determine the rational number $\frac{m}{n}$. At the same time, this is also the result of the division of these numbers: $m : n = \frac{m}{n}$. This use of fraction notation is extended to all rational numbers as another sign of division.

Another Symbol for Division

$$\frac{a}{b} = a : b$$

We still call this notation a fraction, but now both in its numerator and denominator there may be fractions. Instead of $\frac{9}{4} : \frac{21}{10}$, now we write:

$$\frac{\dfrac{9}{4}}{\dfrac{21}{10}}$$

Such a notation is called a **complex fraction**. Based on the definition of division we calculate:

$$\frac{\dfrac{9}{4}}{\dfrac{21}{10}} = \frac{9}{4} : \frac{21}{10} = \frac{\cancel{9}^3}{\cancel{4}^2} \cdot \frac{\cancel{10}^5}{\cancel{21}^7} = \frac{15}{14}$$

This example shows the procedure of calculating a complex fraction. Apart from being able to cancel a particular fraction, we can cancel both the numerator of one fraction with the numerator of the other, as well as the denominator of one fraction with the denominator of the other:

$$\frac{\dfrac{\cancel{9}^3}{\cancel{4}^2}}{\dfrac{\cancel{10}^5}{\cancel{21}^7}} = \frac{\dfrac{3}{2}}{\dfrac{7}{5}}$$

We calculate it so that the product of the external numbers gives the numerator, and the product of the internal numbers gives the denominator of the result:

$$\frac{\dfrac{3}{2}}{\dfrac{7}{5}} = \frac{3 \cdot 5}{2 \cdot 7}$$

How to Calculate a Complex Fraction

$$\frac{\dfrac{a}{b}}{\dfrac{c}{d}} = \frac{a \cdot d}{b \cdot c}$$

Example 3.7.2. Let us calculate:

1. $\dfrac{\dfrac{12}{14}}{\dfrac{9}{21}}$

2. $\dfrac{\dfrac{3}{4}}{\dfrac{9}{10}} : \dfrac{\dfrac{3}{2}}{\dfrac{2}{20}}$

Solution.

1. $\dfrac{\dfrac{\cancel{12}^{6}}{\cancel{14}^{7}}}{\dfrac{\cancel{9}^{3}}{\cancel{21}^{7}}} = \dfrac{\dfrac{\cancel{6}^{2}}{7}}{\dfrac{\cancel{3}}{7}} = \dfrac{2}{1} = 2$　　　2. $\dfrac{\dfrac{3}{4}}{\dfrac{9}{10}} : \dfrac{\dfrac{3}{2}}{\dfrac{2}{20}} = \dfrac{\dfrac{3}{2}}{\dfrac{1}{5}} : \dfrac{\dfrac{3}{1}}{\dfrac{1}{5}} = \dfrac{5}{6} : \dfrac{15}{1} = \dfrac{5}{6} \cdot \dfrac{1}{15} = \dfrac{1}{18}$　□

Example 3.7.3. Applying the formula for total resistance R of resistors R_1 and R_2 connected in parallel:

$$\frac{1}{R} = \frac{1}{R_1} + \frac{1}{R_2}$$

let us calculate the total resistance of the connection in parallel resistances of $\dfrac{3}{10}$ Ohm and $\dfrac{4}{15}$ Ohm.

Solution. Let us insert the known values into the formula and let us calculate:

$$\frac{1}{R} = \frac{1}{\dfrac{3}{10}} + \frac{1}{\dfrac{4}{15}} + \frac{15}{4} = \frac{40+45}{12} = \frac{85}{12}$$

Thus, $R = \dfrac{12}{85}$ Ohm. $\qquad\qquad\qquad\qquad\qquad\qquad\qquad\qquad\qquad\qquad$ $\square$

All the formulas that were valid for standard fractions $\dfrac{m}{n}$, where m is an integer, and n is a positive integer, are valid also for arbitrary fractions $\dfrac{a}{b}$, where a and b are arbitrary rational numbers provided $b \neq 0$.

Some Additional Properties of Multiplication and Division

1. *The reciprocal number of the non-zero number a is the number* $\dfrac{1}{a}$ **(reciprocal number)**

2. $\dfrac{a}{-b} = \dfrac{-a}{b} = -\dfrac{a}{b}$ **(the sign *"walks"* through a fraction)**

3. $\dfrac{a}{b} = \dfrac{a \cdot c}{b \cdot c} = \dfrac{a : c}{b : c}$ **(cancellation and expansion of fraction)**

4. *For $b > 0$, $\dfrac{a}{b} < \dfrac{c}{b}$ precisely when $a < c$*

   For $b < 0$, $\dfrac{a}{b} < \dfrac{c}{b}$ precisely when $a > c$ **(comparison)**

5. $\dfrac{a}{b} + \dfrac{c}{b} = \dfrac{a+c}{b} \qquad \dfrac{a}{b} - \dfrac{c}{b} = \dfrac{a-c}{b}$ **(addition and subtraction)**

6. $\dfrac{a}{b} \cdot \dfrac{c}{d} = \dfrac{ac}{bd}$ **(multiplication)**

7. $\dfrac{a}{b} : \dfrac{c}{d} = \dfrac{a}{b} \cdot \dfrac{d}{c}$ **(division)**

These properties are easy to prove based on the already known properties of rational numbers and the fact that division and multiplication are inverse operations. We must first prove the first property, since division is reduced to multiplication by a reciprocal number. Let $a = \dfrac{m}{n}$. Then

$$\frac{1}{a} = \frac{1}{\dfrac{m}{n}} = \frac{n}{m}$$

so that it really is the reciprocal number of the number a. Further proofs follow easily. For instance, for the fifth property:

$$\frac{a}{b} + \frac{c}{b} = a \cdot \frac{1}{b} + c \cdot \frac{1}{b} = (a+c) \cdot \frac{1}{b} = \frac{a+c}{b}$$

Example 3.7.4. Let us simplify:

1. $\dfrac{1}{2a^2} + \dfrac{1}{6ab^2} + \dfrac{1}{3b^3}$ 2. $\dfrac{3a^2 b}{4c^2 d} \cdot \dfrac{6cd^2}{9ab^2}$ 3. $\dfrac{\dfrac{2a}{3b^2}}{\dfrac{4a}{6b}}$

Solution.

1. $\dfrac{1}{2a^2} + \dfrac{1}{6ab^2} + \dfrac{1}{3b^3} = \dfrac{3b^3 + ab + 2a^2}{6a^2 b^3}$

2. $\dfrac{3a^2\, b}{4^2 c^2\, d} \cdot \dfrac{6^3\, cd^2}{9^3\, ab^2} = \dfrac{ad}{2cb}$

3. $\dfrac{\dfrac{2\, a}{3b^2}}{\dfrac{4^2\, a}{6^2\, b}} = \dfrac{\dfrac{1}{b}}{\dfrac{1}{1}} = \dfrac{1}{b}$

Example 3.7.5. Let us prove that from the formula for resistors connected in parallel

$$\frac{1}{R} = \frac{1}{R_1} + \frac{1}{R_2}$$

it follows that the total resistance can also be calculated according to this formula:

$$R = \frac{R_1 \cdot R_2}{R_1 + R_2}$$

Let us also prove that the total resistance of resistors connected in parallel is smaller than their individual resistances:

$$R < R_1 \quad R < R_2$$

Solution. The formula will now be transformed in the order in which previously the total resistance was calculated:

$$\frac{1}{R} = \frac{1}{R_1} + \frac{1}{R_2} = \frac{R_2 + R_1}{R_1 \cdot R_2}$$

If we look at the reciprocal numbers, we eventually obtain

$$R = \frac{R_1 \cdot R_2}{R_1 + R_2}$$

I leave it to you to explain this chain of inference:

$$\frac{1}{R} = \frac{1}{R_1} + \frac{1}{R_2} \;\rightarrow\; \frac{1}{R} > \frac{1}{R_1} \;\rightarrow\; R < R_1$$

Analogously it is shown that $R < R_2$. $\square$

This example should once again convince us how much more significant it is to know how to work with open descriptions than with closed ones:

The Importance of Open Descriptions

Open descriptions make it possible to discover general laws about numbers, and when these numbers express relations in reality then they also make it possible to discover general laws about reality.

3.8 Equations, Equalities and the Laws of Nature

We have seen that in the set of rational numbers all the **basic arithmetic operations**: addition, subtraction, multiplication and division (except for division by zero), are always feasible. The good properties of these operations give us the freedom and simplicity in calculation that we did not have either in the set of natural numbers nor in the set of integers. Now we will use these advantages in solving equations.

Example 3.8.1. After a hard business day, pirate Giorgio decided to relax a bit. He took several gold coins and parked his boat in front of the *Double Luck* casino. The admission cost five gold coins, but in the casino he was lucky and doubled the number of gold coins that he had on him. He paid five gold coins for parking, and left on his boat for another casino *Free Admission*, where he again doubled his amount of gold coins. However, after he had left and paid six gold coins for parking, he had no coins left, and he wondered why he had not one gold coin in his pocket when he had been winning constantly. How many coins did the pirate Giorgio take when he set out to have fun?

Solution. Let us call the required number of unknown gold coins x. In order to determine how much x is, we have to make use of the information mentioned. After paying the admission to the first casino, Giorgio had $x - 5$ gold coins. Since in the games of chance he doubled this number, he left the casino with $(x - 5) \cdot 2$ gold coins. However, he had to pay 5 gold coins for the parking, so he entered the second casino (where the admission was free)

with $(x-5)\cdot 2-5$ gold coins. He had luck there as well, and so he left with double the number of gold coins, which means $[(x-5)\cdot 2-5]\cdot 2$ gold coins. After paying 6 gold coins for the parking, he had nothing left. Thus,

$$[(x-5)\cdot 2-5]\cdot 2-6=0$$

This is the required information about the unknown number x.

Equation with One Unknown

This type of information about an unknown number, a condition in the form of equality, is called an **equation** (with one unknown).

The solution of the equation is any number which fulfils this condition.

To solve an equation means to find all its solutions.

We will deal in more detail with equations and their application in solving actual problems (like the one with the pirate) in Circle 2. For the moment, we will consider the equation as information about an unknown number x, on the basis of which we will try to discover it. However, unlike in the detective novels by Agatha Christie, we will not need any inspired reasoning by detective Hercule Poirot, but rather simple calculation. The aim is, namely, to obtain from the initial information $[(x-5)\cdot 2-5]\cdot 2-6=0$ the information in the form

$$x = \text{something known}$$

This will be realised by a number of simple inferences. Every such inference consists of a simple step – we will add or subtract the same number both from the left and the right side of equation, or we will multiply or divide them by the same non-zero number. Every such step is possible because these operations are feasible in the set of rational numbers. Every one is correct since, by applying the same operation equally on the left and right sides, we will again obtain equal left and right sides. For instance, we can add to the left and right side of the initial equation the number 100. Briefly, we say that we added 100 to the equation and we write down:

$$[(x-5)\cdot 2-5]\cdot 2-6=0 \quad |+100$$

When we add 100 both to the left and the right side, we get

$$[(x-5)\cdot 2-5]\cdot 2-6+100 = 0+100$$

After simplifying both sides, the equation is

$$[(x-5)\cdot 2-5]\cdot 2+94 = 100$$

Since the step is correct, we have again obtained a true item of information about x. However, the problem is that it is no better than the initial information. The number x is no more known now than before. However, if instead of 100 we add 6 to the equation, we will get better information about the number x, because with this action we will cancel the subtraction of the number 6 on the left side of the equation:

$$[(x-5)\cdot 2-5]\cdot 2-6 = 0 \quad |+6$$
$$[(x-5)\cdot 2-5]\cdot 2-6+6 = 0+6$$
$$[(x-5)\cdot 2-5]\cdot 2 = 6$$

Now we are closer to the number x. Its discovery may be compared to a rabbit in the forest caught in a trap that you want to save by opening the trap. You disassemble the trap from the outside.

We have already cancelled the subtraction of the number 6 by adding the number 6 to the equation. Now we need to cancel the multiplication by the number 2. We will do this by dividing the equation by the number 2:

$$[(x-5)\cdot 2 - 5]\cdot 2 = 6 \quad |:2$$
$$[(x-5)\cdot 2 - 5]\cdot 2 : 2 = 6 : 2$$
$$(x-5)\cdot 2 - 5 = 3$$

The strategy for discovering the number x is clear. Although we can perform arbitrary operations with the equation, we will perform the one that will cancel the last operation done with number x. And this is its inverse operation. Thus we will now cancel the subtraction of number 5 by adding the number 5, and so on:

$$(x-5)\cdot 2 - 5 = 3 \quad |+5$$
$$(x-5)\cdot 2 - 5 + 5 = 3 + 5$$
$$(x-5)\cdot 2 = 8 \quad |:2$$
$$x - 5 = 4 \quad |+5$$
$$x = 9$$

The number x has been discovered! Thus, pirate Giorgio had 9 gold coins in his pocket. Although in every casino he doubled the number of gold coins, he had to pay too much for admissions and parking. Oh my dear Giorgio, there is always a bigger fish. $\square$

We could eliminate all operations precisely owing to the fact that in the set of rational numbers, for every basic arithmetic operation there is its inverse operation, except for multiplication by zero. Not being able to divide by zero is not a restriction in solving equations, since such a situation is very rare (we will deal with this in *Circle 2*). True, we can multiply by zero, but this is completely non-informative. Indeed, if we multiplied the initial equation (or any other) by zero, since multiplication by zero yields zero, we would get $0 = 0$. Therefore, we would not find out anything new, and particularly not about the sought unknown. Simply, by multiplying the equation by zero, we would destroy the information.

Bearing in mind the previous example, let us analyse the process of solving an equation. The permitted steps in solving are:

- *Any description on any side can be replaced by a different description of the same number.* For instance, the description $x + 1 + 2x - 3$ can be replaced by the description $3x - 2$.

- *With both sides we can do the same*, because if we perform the same operation f to both sides of the equality $L = D$, we will obtain equality again: $f(L) = f(D)$. For instance, if $2x - 3 = x + 1$, if we subtract on both sides x, we will get the equality $2x - 3 - x = x + 1 - x$.

These rules are correct and simple, although there are some details that we will leave for *Circle 2*. For the moment, these details will not be significant to us since the only operations that we will do will be addition and subtraction, as well as multiplication and division by a non-zero number. Since we are working in the set of rational numbers, these operations are always feasible. Since every operation has its inverse operation, whenever - by applying one of these operations to one equation - we get another equation, by inverse operation from the second equation we obtain the first one once again. This establishes one significant connection between these equations: *equations are* **equivalent**, *that is to say they have the same solutions*. Generally, for two conditions we say that they are **equivalent** when the same objects satisfy them. *Thus, solving of the equation is reduced to repeated transformations according to the mentioned rules into equivalent equations, until the final equation is obtained whose solution we can "see".* In the previous example such a final equation was the equation $x = 9$. The only number that satisfies this equation is 9. Since the equation $x = 9$ is equivalent to the initial equation from the example, $[(x - 5) \cdot 2 - 5] \cdot 2 - 6 = 0$, the number 9 is at the same time the only solution of the initial equation. Naturally, in transforming the equations we use the permitted steps with a certain strategy, in order to obtain a simple equation by which we see the solution. The basic principle that we use is the **principle of inverse operation**: *when we want to cancel the last operation on one side of the equation, we apply the inverse operation on the equation.*

> ### Procedure (Restricted) for Solving Equations
>
> Our aim is to get an equation which is equivalent to the initial equation and which gives us a solution. We try to get it by progressively applying the following steps to the initial equation:
>
> - Any description on either side can be replaced by a different description of the same number.
>
> - We may add to or subtract from both the left and the right side of the equation the same number. We may also multiply or divide them with the same non-zero number.
>
> One of the basic strategies of the procedure is the principle of inverse operation: in order to eliminate the last operation performed on one side of the equation, we apply both to the left and the right side of the equation its inverse operation.

Example 3.8.2. Pirate Giorgio, angry at being robbed so dishonourably (without a fight), decided to raid the *Double Luck* casino. In the dead of night he managed to get to the cash register and to break the code (by smashing the cash register with an axe several times). With a full bag of gold coins he started towards the exit, but he bumped into the cashier. The cashier requested two thirds of the treasure to keep quiet. Giorgio gave him two thirds of the treasure and 4 gold coins extra as a tip (a bad habit he acquired by hanging out with waiters). At the very exit of the casino he bumped into the security guard and to prevent him from betraying him, he gave the guard half of the rest of the loot and 2 more gold coins. Just as he was about to board his ship, he bumped into the local gendarme and had to give him three fourths of the rest of the treasure (something like a theft tax) and 8 gold coins more as a tip. When he arrived home, he was disappointed to find that he only had one gold coin left. How many stolen gold coins was Giorgio sorry about?

Solution. Having given $\dfrac{2}{3}$ of the robbed treasure x to the cashier, he had $\dfrac{1}{3}$ of the treasure left, and this amount was also reduced by the 4 gold coin tip. Thus, he had $\dfrac{1}{3}x - 4$ gold coins left. In the same way we can determine that after the encounter with the security guard he had $\dfrac{1}{2}\left(\dfrac{1}{3}x - 4\right) - 2$ gold coins, and after meeting the gendarme $\dfrac{1}{4}\left[\dfrac{1}{2}\left(\dfrac{1}{3}x - 4\right) - 2\right] - 8$. Since only one gold coin remained, the required equation for x is:

$$\frac{1}{4}\left[\frac{1}{2}\left(\frac{1}{3}x-4\right)-2\right]-8=1$$

This equation will be solved in the manner already described:

$$\frac{1}{4}\left[\frac{1}{2}\left(\frac{1}{3}x-4\right)-2\right]-8=1 \mid +8$$

$$\frac{1}{4}\left[\frac{1}{2}\left(\frac{1}{3}x-4\right)-2\right]=9 \mid \cdot 4$$

$$\frac{1}{2}\left(\frac{1}{3}x-4\right)-2=36 \mid +2$$

$$\frac{1}{2}\left(\frac{1}{3}x-4\right)=38 \mid \cdot 2$$

$$\frac{1}{3}x-4=76 \mid +4$$

$$\frac{1}{3}x=80 \mid \cdot 3$$

$$x=240$$

If the unknown in the equation can be found in several places, we will keep simplifying the left and the right side of the equation and move the members from one side to the other in order to have the unknown only in one place. The following examples show a typical procedure of solving such equations.

Example 3.8.3. Let us solve:

1. $3 - (2x - 3) = 4(x - 1) - [1 - 2(x - 1)]$

2. $\dfrac{1 + 2(4 - x)}{2} - \dfrac{2 - x}{3} = \dfrac{6 - 2x}{4} + \dfrac{5}{2}$

Solution.

1. First, we will simplify each side:

$$3 - (2x - 3) = 4(x - 1) - [1 - 2(x - 1)]$$
$$3 - 2x + 3 = 4x - 4 - [1 - 2x + 2]$$
$$6 - 2x = 4x - 4 - 1 + 2x - 2$$
$$6 - 2x = 6x - 7 \quad | - 6x - 6$$

Parts with the unknown are moved to one side and the rest to the other:

$$-2x - 6x = -7 - 6$$
$$-8x = -13 \quad | : (-8)$$

By dividing by -8 we will bring x into the clear:

$$x = \frac{13}{8}$$

2. We get rid of the denominator by multiplying the equation by the least common multiple of all the denominators:

$$\frac{1 + 2(4 - x)}{2} - \frac{2 - x}{3} = \frac{6 - 2x}{4} + \frac{5}{2} \quad | \cdot 12$$
$$\frac{1 + 2(4 - x)}{\cancel{2}} \cdot \cancel{12}^{6} - \frac{2 - x}{\cancel{3}} \cdot \cancel{12}^{4} = \frac{6 - 2x}{\cancel{4}} \cdot \cancel{12}^{3} + \frac{5}{\cancel{2}} \cdot \cancel{12}^{6}$$
$$6 \cdot (1 + 2(4 - x)) - 4 \cdot (2 - x) = 3 \cdot (6 - 2x) + 5 \cdot 6$$

We simplify both sides:

$$6 \cdot (1 + 8 - 2x) - 8 + 4x = 18 - 6x + 30$$
$$6 + 48 - 12x - 8 + 4x = 18 - 6x + 30$$
$$-8x + 46 = 48 - 6x \quad | + 6x - 46$$

We transfer the parts:

$$-8x + 6x = 48 - 46$$
$$-2x = 2 \quad | : (-2)$$
$$x = -1$$

The obtained solution can be always checked by substituting it in the equation. By including $x = -1$ into the equation we get

$$\frac{1 + 2(4 - (-1))}{2} - \frac{2 - (-1)}{3} = \frac{6 - 2 \cdot (-1)}{4} + \frac{5}{2}$$

The calculation yields $\frac{9}{2} = \frac{9}{2}$. We have obtained a true statement. Thus, the number -1 satisfies the condition and it really is the solution of the equation. By verifying the obtained solution we can discover the error in solving the equation. There are also other reasons that require verification from time to time, but we will talk about it later, when we study equations in more detail in *Circle 2*.   □

In the software *SageMath* we have the command *solve* for solving equations. The command *solve* has two inputs – we have to specify which equation we want to solve and according to which variable. In this way, we can solve the last equation from the previous example:

```
solve((1+2*(4-x))/2-(2-x)/3==(6-2*x)/4+5/2,x)
# The sign for equality in an equation is written as ==!
```

We will get

```
[x == -1]
```

Example 3.8.4. Since he was left without any cash, pirate Giorgio remembered the treasure map that his late father Domnius had left him. Following the instructions on the back of the map, he sailed to the island of *Fumija* and found a stone with an engraved dead head. When he lifted the stone, he was struck by a current. He did not notice that underneath the dead head there was the inscription *HIGH VOLTAGE! DANGER!*. Thanks to the electric shock he realised that he had made a mistake and soon he found the right stone with its dead head. When he lifted it, underneath he found on impregnated leather the rest of the instructions. They said *"From this place start walking towards the East and make double as many steps as you will make later towards the South, and a hundred steps more. When you make a total of five times more steps than what you made when walking towards the South, the treasure will be under your feet. Love from your dad Domnius. "* After a good cry in remembrance of his dear dad, who was much respected by everyone as a diligent and pious pirate, he started to follow the instructions devotedly. But alas, he could by no means determine how far in which direction he had to go. Finally, he sat down crestfallen on the stone and remembered again his late dad. And it was only then that he understood why his father kept constantly warning him he should learn math, and who was now to blame that he did not listen to his dad? If you happen to arrive on the island of *Fumija* (I will not tell you where this island is so that you will not arrive there before me) and find the engraved dead head (that does not warn of high voltage), how many steps will you take eastwards, and how many southwards in order to find the treasure?

Solution. If we translate this problem into math language, we will get a simple equation. Let x be the number of steps that need to be taken southwards. It follows from the instructions that one should go $2x + 100$ steps eastwards. The instructions also say that the total number of steps eastwards and southwards has to be 5 times greater than x. Hence,

$$x + 2x + 100 = 5x$$

We will solve the equation easily in the manner already described:

$$x + 2x - 5x = -100 \;\to\; -2x = -100/:(-2) \;\to\; x = 50$$

Thus, one should take $2x + 100 = 200$ steps eastwards, and then take $x = 50$ more steps southwards. $\qquad\square$

Here is (preventively) a little of math against electric shocks.

Example 3.8.5. Which resistor R_2 has to be connected parallel with resistor $R_1 = 20$ Ohm, if we want the total resistance of the parallel connection to be $R = 10$ Ohm?

Solution. We already know the formula for parallel connection of resistance:

$$\frac{1}{R} = \frac{1}{R_1} + \frac{1}{R_2}$$

It connects the total resistance R with resistances R_1 and R_2 of parallel connected resistors. In the example 3.6.1 on page 137 we applied it for the calculation of total resistance. We introduced into the formula the known resistances R_1 and R_2 and calculated. Nobody now needs to give us a new formula into which we will introduce R and R_1 in order to calculate R_2. Into the existing formula we simply introduce the known values for R and R_1 and we solve the equation:

$$\frac{1}{10} = \frac{1}{20} + \frac{1}{R_2} \;\to\; \frac{1}{10} - \frac{1}{20} = \frac{1}{R_2} \;\to\; \frac{1}{20} = \frac{1}{R_2} \;\to\; R_2 = 20$$

Thus, in order to halve by means of parallel connection the initial resistance of 20 Ohm, the other resistor also has to be 20 Ohm.

Not only is it that nobody has to give us a new formula for calculation, but we can deduce it by ourselves! In the same order in which we have solved the equation, we can express R_2 by means of R and R_1:

$$\frac{1}{R} = \frac{1}{R_1} + \frac{1}{R_2} \;\rightarrow\; \frac{1}{R} - \frac{1}{R_1} = \frac{1}{R_2} \;\rightarrow\; \frac{R_1 - R}{RR_1} = \frac{1}{R_2} \;\rightarrow\; R_2 = \frac{RR_1}{R_1 - R}$$

Once again we can see that *our operations with variables rather than the names of concrete numbers yield much more.* Not only did we solve the concrete problem but we also found the formula with which we can solve all problems of such a type (the total resistance and one resistor are known, and the other resistor needs to be found). $\square$

Using equalities (formulas) we frequently express the laws according to which the quantities in certain situations are related. As an example, a simple law for a simple situation will be introduced now and we will see how it helps us in connecting the quantities in more complex situations as well. When a car moves uniformly at a certain speed v, it will in time t travel the distance s:

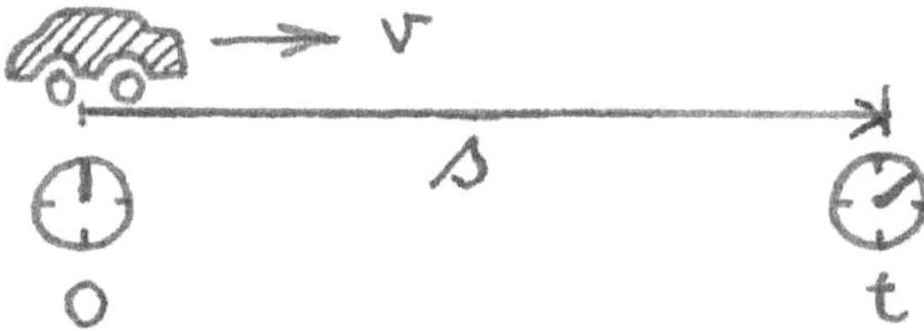

The mentioned values are connected by the following formulas:

$$s = v \cdot t \qquad v = \frac{s}{t} \qquad t = \frac{s}{v}$$

Every formula expresses a value by means of the others. *But all these are variants of one formula, and therefore not everything needs to be remembered.* It is sufficient to remember just one, for example the first:

$$s = v \cdot t$$

If the speed of movement v and the elapsed time t are given, by inserting the values for v and t we will easily calculate the travelled distance. If the elapsed time t and the travelled distance s are known, in order to calculate the speed v, we will insert the known values for s and t into the formula and solve the equation. Or even better, by transforming the formula we will express the speed v by means of the distance s and time t, and then insert the values:

$$s = vt \quad | : t \quad \rightarrow \quad \frac{s}{t} = v \quad \rightarrow \quad v = \frac{s}{t}$$

Example 3.8.6. If a car travelled at a speed of 80 kilometres per hour, what distance did it travel in $\frac{3}{4}$ hours? How much time would a car need to cover this distance by moving at a speed of 120 kilometres per hour?

Solution. Into the formula for the travelled path $s = v \cdot t$ we will insert speed $v = 80$ and time $t = \frac{3}{4}$:

$$s = vt = 80 \cdot \frac{3}{4} = 60$$

Thus, the car travelled 60 kilometres.

In order to solve the second part of the problem, we will express t by means of s and v, and then we will insert $s = 60$ and $v = 120$:

$$s = vt \quad | : v \quad \rightarrow \quad t = \frac{s}{v}$$

Thus, $t = \frac{60}{120} = \frac{1}{2}$ hours.

The driver who was speeding and thus put himself and others in danger, arrived only 15 minutes before the driver who drove at a moderate speed. It is likely that he spent the time he had saved bragging about having arriving quickly. As my friend Joe would say: in traffic the winner is not the one who arrives earlier but the one who arrives at all.

The second part of the task could be solved by introducing into the initial equation the known values for s and v, and then by solving the equation. However, it is always better to express the required quantity by means of the known ones (by solving the equation in the general manner), and only then to substitute the values. This is how we develop the very powerful ability of thinking with variables and discovering new relations between quantities. $\square$

> ### Procedure of Finding an Unknown Value
>
> When an equality connects some quantities one of which is unknown, the unknown quantity can be found in the following way. We insert into the equality values for the known quantities. We will get an equation and by solving it we will find the unknown quantity. However, it is better to express the unknown quantity by means of the known quantities, using the same procedure which is used to solve the equation, and only then to insert values for the known quantities. In this way we get not only the solution of the concrete problem but also a new equality that solves all the problems of this type.

We will now apply the formula for uniform motion in the analysis of more complex situations.

Example 3.8.7.

1. At the foot of the mountain Romeo was calling after Juliet, but the only answer he got was his echo, and this with $3\frac{1}{2}$ seconds delay. If sound travels at a speed of 330 metres per second, how far were the mountain rocks from Romeo?

2. A train 70 metres long was passing over a bridge at a speed of 20 metres per second. Romeo was bored under the bridge waiting for Juliet. He discovered that the bridge kept vibrating for a full 6 seconds. He was bothered by the questions: Where is Juliet? How long is the bridge? When will Juliet come?

3. When he saw Juliet, Romeo started to run fast and in 2 seconds he had already reached a speed of 6 m per second. What distance did he cover?

Solution.

1. Let us call the unknown distance to the rocks s. Since the speed of sound is $v = 330$ metres per second, and the sound travel time is $t = 3\frac{1}{2} = \frac{7}{2}$ seconds, somebody could be misled by these symbols to make the wrong conclusion that according to the formula for uniform motion $s = v \cdot t$. However, this is not so. *Formulas cannot be formally applied. In order to be able to apply a formula correctly, we have to understand its content. We have to understand the meaning of the symbols*

it contains, that is, the values that it connects. The formula for uniform motion connects the movement speed v, the passed time t and the distance covered during this period of time. In this case this distance is not s but $2s$, since during this period of time the sound travelled to the rocks and returned back to Romeo. Thus, it travelled a distance of 2s. *Only by understanding the formula for uniform motion can we relate correctly the unknown value s with the known values v and t*:

$$2s = v \cdot t$$

It is simple from this equation, dividing by 2, to calculate the required distance from the rocks:

$$s = \frac{1}{2}vt = \frac{1}{2} \cdot 330 \cdot \frac{7}{2} = 577\frac{1}{2} \text{ metres}$$

2. Where Juliet was and when she would arrive were things known only to her. We can help Romeo only regarding the length of the bridge. Formulas are very efficient, but they are not omnipotent. In order to solve the question of the bridge length we will set adequate equations. *Everything spoken about has to be named. Even the known values should be assigned symbolic markings because in this way our thinking is more general and clearer.* Therefore, we will call the train length l, and the unknown length of the bridge d. Travelling at a speed of v, during the bridge's vibration the train travelled the distance of $l + d$.

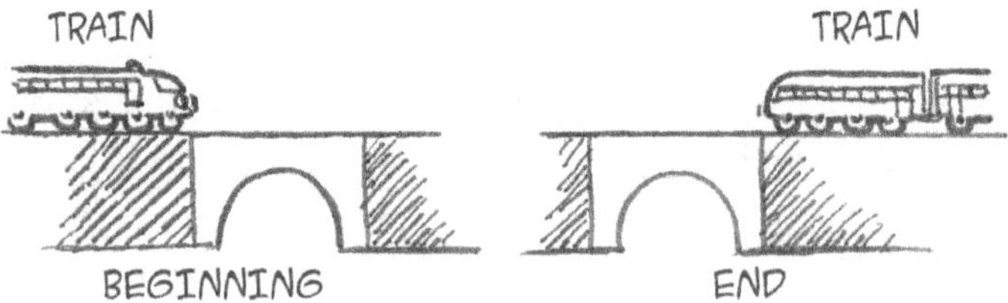

Therefore, according to the formula for uniform motion, the relation among the mentioned values is as follows:

$$l + d = v \cdot t$$

When we subtract l from both sides of the equation, the unknown value d will be expressed by means of the remaining known values:

$$d = vt - l = 20 \cdot 6 - 70 = 50 \text{ metres}$$

3. We cannot obtain s here according to the formula $s = v \cdot t = 6 \cdot 2 = 12$ metres. The formula $s = v \cdot t$ is valid, of course, only for uniform motion, and we are talking here about accelerated motion. Thus, the formula $s = v \cdot t$ cannot be applied here. If we assume that Romeo's speed was increasing uniformly, then we can use the formulas for uniform accelerated motion. It is known from physics that in such a situation the travelled distance s depends on the passed time t, speed v_0 at the beginning, and speed v at the end of the passed time t, according to the next formula:

$$s = \frac{v_0 + v}{2} t$$

Since in our case the speed at the beginning was zero, it follows that

$$s = \frac{0 + 6}{2} \cdot 2 = 6 \text{ metres}$$

$\square$

How the Formulas are Applied

We often link the unknown value with the known ones by applying a formula (equality) which expresses the law of the given situation. In order to apply the formula correctly, we have to understand it. We have to know in which situations it is applicable and which quantities it links.

The *SageMath* command *solve* also helps us to express one quantity by means of others from the given equality. For instance, from the formula for the resistance of the connection in parallel $\dfrac{1}{R} = \dfrac{1}{R_1} + \dfrac{1}{R_2}$ we can express the resistance R_2 by means of other resistances:

```
var('R,R1,R2')
show(solve(1/R==1/R1+1/R2,R2))
```

We will obtain

$$R_2 = -\frac{RR_1}{R-R_1}$$

Let us return to solving equations. The basic rule in solving equations is simple and understandable: *apply the same operation to both sides of the equation.* On this basis we can deduce other rules which are in certain situations simpler. Let us solve, for example, the equation $3x + 6 = 15$:

$$3x + 6 = 15 \quad |-6 \quad \rightarrow \quad 3x = 15 - 6 \quad \rightarrow \quad 3x = 9 \quad |:3$$

$$\rightarrow \quad x = \frac{9}{3} \quad \rightarrow \quad x = 3$$

The total effect of subtracting the number 6 on both sides of the equation is that on the left side the adding of the number 6 is eliminated, and on the right side the subtracting of the number 6 has appeared. The total effect of division by the number 3 is that on the left side the multiplication by the number 3 is eliminated, and on the right side the division by the number 3 has appeared. Thus, you can convince yourself of the correctness of the following rules, and I believe you know them from school.

Moving Descriptions from One Side of the Equation to the Other

- The description which on the one side of the equation is added to the remaining part can be moved to the other side, where it is subtracted from that side, and vice versa.

- The description of a non-zero number which on the one side of the equation is multiplied with the remaining part can be moved to the other side using it for the division of that side, and vice versa.

Example 3.8.8. Let us solve the following equations by the rules of transposing:

1. $2x - 3 = 4x - 5$ 2. $\dfrac{3x}{4} = \dfrac{5}{6}$

Solution.

1. $2x - 3 = 4x - 5 \quad \rightarrow \quad 2x - 4x = -5 + 3 \quad \rightarrow \quad -2x = -2 \quad \rightarrow$

   $x = \dfrac{-2}{-2} \quad \rightarrow \quad x = 1$

2. $\dfrac{3x}{4} = \dfrac{5}{6} \quad \rightarrow \quad x = \dfrac{5 \cdot 4}{6 \cdot 3} \quad \rightarrow \quad x = \dfrac{10}{9}$ $\qquad\qquad\square$

When we link the known and unknown quantities, it may happen that we have several unknown quantities and several pieces of information about them. Then the translation is not one equation, but **a system of equations**, several equations with several unknowns. **The solution of a system of equations** (of a system of conditions) with, for example, two unknowns (with two variables) is a pair of numbers which, when taken in order for the values of variables, fulfil (satisfy) all the equations (conditions) of the system. For instance, for the system of equations

$$x + 2y = 4$$
$$x - y = 1$$

the pair (1,2) is not the solution because when instead of x we put the number 1 and instead of y the number 2 we will obtain false statements:

$$1 + 2 \cdot 2 = 4$$
$$1 - 2 = 1$$

In the same way, the pair (0,2) is not the solution either, since, although it satisfies the first equation, it does not satisfy the second one:

$$0 + 2 \cdot 2 = 4$$
$$0 - 2 = 1$$

However, the pair (2,1) is the solution since it satisfies both conditions:

$$2 + 2 \cdot 1 = 4$$
$$2 - 1 = 1$$

In order to make it clearer which number in the pair we have replaced with which variable, we also say that the solution of the system is the pair $x = 2$ and $y = 1$.

Example 3.8.9. In a forest by the road, the cruel Istrian Luciano's gang (Istria is a very peaceful part of Croatia) were relaxing. Along the path the gang of an even more cruel Istrian, Jakob, came along. Respecting the unspoken agreement on sustainable development they did not have a fight but rather a friendly discussion. Luciano remarked *"If you gave us one of your men, we would have the same number of men as you."* Not wanting to lag behind in this intellectual conversation Jakob answered *"If you gave us one man, we would have twice as many men as you."* They did not know that Inspector Clouseau was sitting in the cavity of a tree, and noting everything down carefully. But in the process, he clumsily elbowed a wasps' nest in the tree. The wasps stung him so many times that he could not in any way conclude how many members each gang had. Let us help the clumsy inspector.

Solution. Let us designate the number of people in Luciano's group with L and in Jakob's group with J. Then the given information can be transformed into the following equations:

$$L + 1 = J - 1$$
$$(L - 1) \cdot 2 = J + 1$$

We have obtained two pieces of information about two unknowns. The permitted steps in discovering the unknowns are the same as in the case of one equation, but now we apply them to both equations. The strategy is as follows. We will use one equation in order to describe one unknown by means of the other. For instance, L can be expressed from the first equation by means of J:

$$L = J - 2$$

Thus, we have used the first piece of information to describe L by means of J. In this way we have reduced the problem to searching for J. When we determine J, from the above description we will easily calculate L. In order to find J, we will use the second item of information. True, it talks also about L and about J, but now we know how to express L by means of J. When in the second piece of information L is replaced by its description by means of J, $L = J - 2$, it will talk only about J:

$$(J - 2 - 1) \cdot 2 = J + 1$$

Thus, we have reduced the problem of solving two equations with two unknowns to the problem of solving one equation with one unknown. Formally, we achieved this by replacing L with the description by means of J, and this method of solving the system is called the **substitution method**. Now we solve the equation with one unknown. We easily get that $J = 7$. Knowing J, from the description of L by means of J we will calculate also L: $L = J - 2 = 7 - 2 = 5$.

Dear Inspector Clouseau, Luciano's gang has 5 members, and Jakob's has 7. $\square$

Method of Substitution

The method of substitution is the following procedure for solving a system of equations:

1. We use one equation to express one unknown by means of the remaining ones.

2. In all the other equations we replace this unknown with the obtained expression.

3. We will get fewer equations with fewer unknowns.

4. With the obtained equations we repeat the procedure until we get one equation with one unknown.

5. By solving this equation we will obtain not only its unknown, but, going back to the previous equations, we will obtain all the unknowns.

The method of substitution is the most efficient method of solving a system of equations, but it is not omnipotent. It may occur that in no equation can one unknown be substituted by means of another one, or on the other hand that we obtain an excessively complex description. Then an attempt is made to solve the system in another way – **by combination** of more complex equations into simpler ones. I will illustrate this on the following system of equations:

$$4x + 7y = 6$$
$$6x + 11y = 8$$

This system could be solved by the substitution method, but the procedure would get a little bit complicated (try it). Instead, we will combine the equations into a simpler one. Combining is founded on the logical property of equality that the "same combination of equals yields equal":

$$L_1 = R_1, \; L_2 = R_2 \quad \rightarrow \quad C(L_1, L_2) = C(R_1, R_2)$$

where C is any operation (combination) of two numbers. For instance, if we added the left sides of the equations and the right sides of the equations (then we say that we have added the equations) we would get the equation

$$
\begin{array}{rcl}
4x + 7y & = & 6 \\
6x + 11y & = & 8 \quad \big|+ \\
\hline
4x + 7y + 6x + 11y & = & 6 + 8
\end{array}
$$

The obtained equation is no simpler than the initial ones. However, if there were opposite coefficients of one unknown, we would obtain a simpler equation, since this unknown would disappear from the equation. Therefore, we will first multiply the equations by appropriate numbers so that we get opposite coefficients of one unknown. For instance, in order to obtain opposite coefficients of the unknown x, we will multiply the first equation by -3 and the second by 2 (we will get ± 12 – the smallest common multiple of the existing coefficients 4 and 6):

$$
\begin{aligned}
4x + 7y &= 6 \quad |\cdot(-3) \\
6x + 11y &= 8 \quad |\cdot 2
\end{aligned}
$$

Now that we have obtained opposite coefficients of the unknown, we will add the equations:

$$
\begin{aligned}
-12x - 21y &= -18 \\
12x + 22y &= 16 \qquad \Big|+ \\
\hline
-12x + 12x - 21y + 22y &= -18 + 16
\end{aligned}
$$

We have obtained a simple equation in which one unknown *has disappeared*: $y = -2$. If we introduce this value for y, for instance into the first equation, we will get the corresponding x:

$$
4x + 7\cdot(-2) = 6 \quad \rightarrow \quad 4x = 6 + 14 \quad \rightarrow \quad x = 5
$$

Thus, the solution of the system is the pair $(5, -2)$. This method of solving by combination is called the **method of opposite coefficients**. Naturally, there are also other ways that combinations can be used.

Method of Opposite Coefficients

The system of two linear equations with two unknowns (the equations of the form $ax + by = c$) is best solved by using the **method of opposite coefficients**:

1. We multiply each equation with an appropriate number to obtain opposite coefficients of one of the unknowns.

2. By adding the equations we obtain one equation with one unknown, by which we determine this unknown.

3. We introduce the obtained value of the unknown into one of the initial equations in order to get the respective value of the second unknown.

Example 3.8.10. Let us solve the following systems:

1. $\begin{aligned} -3x + 5y &= 8 \\ x + 7y &= 6 \end{aligned}$

2. $\begin{aligned} 2x - 4y &= 1 \\ 4x - 8y &= 2 \end{aligned}$

3. $\begin{aligned} -3x + 2y &= 7 \\ -6x + 4y &= 7 \end{aligned}$

Solution.

1. We will multiply the second equation by 3 in order to get the opposite coefficients of the unknown x:

$$\begin{array}{rcl} -3x + 5y &=& 8 \\ 3x + 21y &=& 18 \\ \hline 26y &=& 26 \\ \to y &=& 1 \end{array} \quad \Big|+$$

We insert the obtained value for y into the second equation:

$$x + 7 \cdot 1 = 6 \quad \to \quad x = -1$$

The solution is the pair $(-1, 1)$.

2. We will multiply the first equation by -2 in order to get the opposite coefficients of the unknown x:

$$\begin{array}{rcl} -4x + 8y &=& -2 \\ 4x - 8y &=& 2 \\ \hline 0 &=& 0 \end{array} \quad \Big|+$$

When we add the equations, not only does one unknown disappear, but the second one disappears as well and we obtain something that already know: that 0 = 0. The fact that the equations are cancelled by addition means that they are very similar. Indeed, if we were to multiply the first equation by 2 instead of -2, we would obtain precisely the second equation:

$$2x - 4y = 1 / \cdot 2 \quad \rightarrow \quad 4x - 8y = 2$$

This means that the two equations (conditions) are equivalent (have the same solutions), so that one is redundant (it does not provide new information). We can forget it, since its solutions are identical to the solutions of the remaining equation. Thus the solutions are all pairs of numbers that satisfy this condition. If we express x by means of y

$$x = \frac{1}{2} + 2y$$

we can see that whichever number we take instead of y (e.g. $y = 1$) we will get an adequate value for x (for $y = 1$, $x = \frac{5}{2}$). Thus this system has infinite solutions, all pairs of numbers of the form $(\frac{1}{2} + 2y, y)$ where y is an arbitrary number.

3. We will multiply the upper equation by -2 in order to obtain the opposite coefficients of the unknown x:

$$
\begin{array}{rcll}
6x - 4y & = & -14 & \\
-6x + 4y & = & 7 & \Big| + \\
\hline
0 & = & -7 &
\end{array}
$$

Chasing the unknowns, we have obtained a falsehood! However, this can be easily interpreted. Indeed, *transforming from one set of equations to another by permitted steps is in fact, as we have already pointed out, correct conclusion-making.* It shows us that if for some x and y (that we are searching for) the initial equalities are valid, then the final ones are valid as well. If, however, the final equality is false, it means that the initial set of equalities for any x and y fail to be fulfilled either. We also say that the given set of equations (conditions) is unsatisfiable or that it is contradictory. Thus, this system of equations has no solution.   □

The command *solve* in the software *SageMath* also helps us in solving systems of equations. Using it, we will solve, for example, the first system of equations from the previous example, with the following command:

```
var('y')
solve([-3*x+5*y==8,x+7*y==6],x,y)
```

We will get

```
[[x == -1, y == 1]]
```

Example 3.8.11. A messenger pigeon started from town A towards town B flying at a uniform speed of 20 km/h. At the same time (from town B towards town A) there started a messenger dove flying at a speed of 15 km/h. If the towns are at a distance of 7 km, at what point and when will the pigeon meet the dove?

Solution. Let us call the known speeds of the pigeon and the dove $v_1 = 20$ km/h and $v_2 = 15$ km/h, and the distance of the towns $d = 7$ km. The unknown paths that the pigeon and the dove had flown before their encounter will be called s_1 and s_2, and the time until the encounter t. We will apply the formulas for uniform motion and the relationship between the distances to connect the unknown values with the known ones:

$$s_1 = v_1 t \qquad s_2 = v_2 t \qquad d = s_1 + s_2$$

Since the first two equations describe s_1 and s_2 by means of t, we will replace with these descriptions s_1 and s_2 in the third equation and get one equation with one unknown:

$$d = v_1 t + v_2 t \quad \rightarrow \quad d = (v_1 + v_2)t \quad \rightarrow$$

$$t = \frac{d}{v_1 + v_2} = \frac{7\,\text{km}}{20\,\text{km/h} + 15\,\text{km/h}} = \frac{1}{5}\,\text{h} = 12\,\text{min}$$

Thus, the time until encounter is 12 minutes, and the covered paths of the pigeon and the dove are

$$s_1 = v_1 t = 20\,\frac{km}{h} \cdot \frac{1}{5}\,h = 4\,km \qquad s_2 = v_2 t = 15\,\frac{km}{h} \cdot \frac{1}{5}\,h = 3\,km$$

The command *solve* in the software *SageMath* helps us in solving the system of equalities. Thus, from the system of equalities from the previous example we will express the unknown values s_1, s_2 and t by means of the known values v_1, v_2 and d by the following command:

```
var('s1,s2,t,v1,v2,d')
show(solve([s1==v1*t,s2==v2*t,d==s1+s2],s1,s2,t))
```

We will get

$$[[s_1 = \frac{dv_1}{v_1 + v_2}, s_2 = \frac{dv_2}{v_1 + v_2}, t = \frac{d}{v_1 + v_2}]]$$

Let us dwell a little more on the application of equations in solving "actual" problems. A problem described in natural language was translated into mathematical language. In this translation we named all the values we did not know. Every item of information about unknown values was translated into one equation. In this linking of known and unknown values we also used the laws (which are equalities) that "cover" this situation. The translation was a system of equations, a well-defined problem in mathematical language for which we have simple means of solutions that enable us, unlike in natural language, to efficiently find the solutions. However, in the mathematical part we only solved the mathematical problem (found the solutions to the system of equations). This does not necessarily have to be the solution of the actual problem, since it is possible that we made a mistake in translation, used false laws or maybe overlooked some important characteristics of the situation. However, this is no longer a matter of mathematics but rather of knowing the area to which mathematics has been applied. This will be discussed in more detail in *Circle 2*.

The application of equations illustrates a typical method of applying mathematics:

Typical Application of Mathematics

The actual problem, taking into account the laws of the domain of the problem, with some simplifications and by means of the meaning of mathematical notions, is translated from natural language into mathematical language. The translation is a mathematical object (in our case this was a system of equations). We also say that we have obtained a mathematical model of the problem. In mathematical language we usually have efficient formal methods that provide a solution. However, they give the solution to the mathematical problem. This solution needs to be interpreted so as to see whether it is also the solution to the actual problem. It is possible that the translation of the actual problem into the mathematical problem was bad. However, the translation skill is no longer simply a matter of knowing mathematics, but also of knowing the context in which the problem has occurred.

Usually, several laws describe a certain situation, and not just one. We use them in order to obtain new laws or in order to combine in the given situation several unknown values with the known ones. For a concrete example we can take the situation of uniform accelerated motion.

Example 3.8.12. Uniformly accelerated motion is motion under the influence of constant force. As, for instance, when you accelerate uniformly the speed of a car (and neglect the variation in air resistance at the change of speed). Here, the speed v of the body rises uniformly according to the law $v = v_0 + at$ and the travelled path s increases according to law $s = v_0 t + \dfrac{1}{2}at^2$, where v_0 is the initial speed, a is acceleration of the body and t is the time passed.

1. By eliminating t let us find the connection between the remaining quantities.

2. By eliminating a let us find the connection between the remaining quantities.

Solution.

1. Let us express from the equation $v = v_0 + at$ the time t using the remaining values, $t = \dfrac{v - v_0}{a}$, and let us eliminate it from the equation $s = v_0 t + \dfrac{1}{2}at^2$:

$$s = v_0 \left(\frac{v - v_0}{a}\right) + \frac{1}{2}a\left(\frac{v - v_0}{a}\right)^2$$

Applying the rules for equalities, we will get

$$v^2 = v_0{}^2 + 2as$$

2. Let us express from the equation $v = v_0 + at$ the acceleration a using the remaining values, $a = \dfrac{v - v_0}{t}$, and let us eliminate it from the equation $s = v_0 t + \dfrac{1}{2}at^2$:

$$s = v_0 t + \frac{1}{2}\frac{v - v_0}{t}t^2$$

Applying the rules for equalities, we will get

$$s = \frac{v_0 + v}{2} \cdot t$$

Thus, from certain laws that describe a situation, we have derived new laws for that situation!   □

The last example illustrates the most important property of the language of equalities:

The Language of Equalities and Laws of Nature

By measuring, we assign numbers to physical phenomena. These numbers characterise the phenomena. Through these assignments the laws of phenomena often occur as equalities between measured values. Having efficient tools for work with equalities (simplifying each side of the equality and applying the same operation to both sides of the equality) we obtain from one set of equalities some others. Thus we discover new laws in the considered phenomena.

In the *SageMathTutorial* on the web site of the book SageMath is described in more detail how the equations in the software *SageMath* are solved.

You can find more about application of equations in solving "real" problems at `https://en.wikipedia.org/wiki/Word_problem_(mathematics_education)`

3.9 Per cents and vipers

When I was a kid, vipers were a mystery to me in nature, and so were "per cents" in the world of adults. I felt that nature would become my friend if I stopped being afraid of vipers. I thought that I would understand the world of adults if I understood percentages. To understand what inflation of three per cent was or what an autumn discount of fifty per cent was seemed to me like getting a ticket to the world of adults. I have stopped being afraid of vipers. Liberated from fear and understanding vipers, I understand better the whole of nature. However, with percentages it was different. I came to understand the per cents but the world of adults has still remained hard to understand. So many people suffer because of environmental pollution, yet adults are afraid of vipers. Long live the vipers!

Percentages lose their mystical power when they are translated into standard mathematical language. Let us see, for example, what it means if the price of shoes was 240 kuna, and now, on discount, the price is 15 per cent (we write 15%) less. First of all, the expression 15% is only a shortcut in speech for a fraction with enumerator equal to 100:

15% = 15 per cent (in Latin, cent means one hundred) =

15 per hundred = 15/100

Thus, we can express the discount in standard mathematical language:

$$15 \text{ per cent of } 240 = \frac{15}{100} \cdot 240$$

The translation is a simple description which is easy to calculate:

$$\frac{15}{100} \cdot 240 = 36$$

So, the new price is 240–36 = 204 kuna.

And that is all. However, the language of percentages is often used for cheating, so we will dwell upon it a little more.

Usually, percentages are used to express a part P of some positive quantity Q. For example, when the price $Q = 240$ kuna has been changed to 206 kuna the absolute value (the amount) of change is $P = 36$ kuna. The relationship between part or change P and whole quantity Q is expressed by their proportion:

$$k = \frac{P}{Q}$$

So, in our example: $k = \dfrac{36}{240} = \dfrac{3}{20}$

If we multiply the previous formula by Q we get

$$P = k \cdot Q$$

So, the proportion k is the number by which the quantity Q has to be multiplied to get its part P. In our example: $P = \dfrac{3}{20} \cdot 240 = 36$

And where is the per cent? When we express the proportion k as a fraction with the numerator 100 we call it *per cent*:

$$k = \frac{p}{100}$$

In our example: $k = \dfrac{3}{20} = \dfrac{3 \cdot 5}{20 \cdot 5} = \dfrac{15}{100}.$

Formulas for Vipers (sorry, Per cents)

The proportion of part P in relationship to the quantity Q is the number

$$k = \frac{P}{Q}$$

So, the following formula gives the part P from the quantity Q:

$$P = k \cdot Q$$

When we express the proportion k as a fraction with the numerator 100 we call it **per cent**:

$$k = \frac{p}{100} = p\%$$

In this notation the formulas become:

$$\frac{p}{100} = \frac{P}{Q} \quad \text{i.e.} \quad P = \frac{p}{100} \cdot Q$$

Example 3.9.1.

1. Because of workers' complaints about low salaries insufficient for normal life, the company management decided to raise the salaries of all employees by 10%. Is it fair? To give a proper answer let us calculate how much the rise in the manager's salary of 15 000 kuna (about 2000 euro) is, and how much the rise in the workers' salary of 2300 kuna (about 300 euro) is.

2. If the price of an urban transport ticket is raised from 5 kuna to 7 kuna (about one euro), what is the percentage of the rise? Is it too much? If a mother and father have three children and their monthly urban transport expenses are 450 kuna (about 60 euro) what will their expenses be after the rise? □

Solution.

1. Although the rise in salary is the same for all when it is expressed in percentages, the real rises are not the same for all, because a percentage of a bigger number is a bigger number. The salary of the manager will rise $\frac{10}{100} \cdot 15\,000 = 1500$ kuna (about 200 euro) and the salary of the workers will rise $\frac{10}{100} \cdot 2300 = 230$ kuna (about 30 euro). The manager will say that it is fair because his work is worth more. The worker will say that it is not fair because the salary of the manager has been raised much more than his salary although the manager's standard of living was not threatened. Moreover, after this rise the workers' standard of living will not be significantly better. However, if the manager's salary stayed the same, the workers could get a better rise. What is more just?

2. Although we could solve the problem directly, we will use formulas for percentages:

$$\frac{p}{100} = \frac{P}{Q} \quad \text{i.e.} \quad P = \frac{p}{100} \times Q$$

As with other formulas, we must understand these formulas, if we want to apply them correctly. We must know the meaning of the symbols in the formulas. In this case, we must know that Q is some initial quantity, P is its part or change, and $\frac{p}{100}$ is the proportion of P and Q expressed as a percentage. Now, we can apply formulas. In our example $Q = 5$, and $P = 2$, so

$$\frac{p}{100} = \frac{P}{Q} = \frac{2}{5} = \frac{2 \times 20}{5 \times 20} = \frac{40}{100} = 40\%$$

Applying the same formulas, we will find the new family expenses. The initial expenses will be denoted Q_1 and not Q, because Q is a symbol for something else in this example. *It is a good practice to denote different quantities with different symbols.* The unknown rise P_1 will be calculated applying the formula which connects it with the known quantities:

$$P_1 = \frac{p}{100} \cdot Q_1 = \frac{40}{100} \cdot 450 = 180 \text{ kuna}$$

The family must ensure 180 kuna more (about 25 euro), that is to say their monthly urban transport expenses will be $450 + 180 = 630$ kuna (about 85 euro). Although it might seem that the increase in the ticket price from 5 to 7 kuna (a rise of about 25 cents) was not too much, with bigger amounts it could be very significant. □

How do we find percentages if we cannot express a rational number q as a fraction with numerator 100? Then we can express it as a double fraction with the numerator 100:

$$a = \frac{a \cdot 100}{100}$$

For example:

$$\frac{1}{3} = \frac{\dfrac{1}{3} \cdot 100}{100} = \frac{\dfrac{100}{3}}{100} = \frac{100}{3}\%$$

In the next chapter about real numbers, we will see how decimal notation gives an easy way to get (approximate) percentages.

And that is it with the percentages. One hundred per cent!

3.10 Powers

The set of rational numbers enables also simpler work with powers. In the chapter on natural numbers we described what a power a^b for natural numbers a and b (where 0^0 is not defined) is, where a is called the basis of the power and b is the power exponent. Thus, for instance

$2^3 = 2 \cdot 2 \cdot 2 = 8$, whereas

$2^0 = 1.$

In the chapter about integers we allowed that the base of a power be a negative number. For instance,

$$(-2)^3 = (-2) \cdot (-2) \cdot (-2) = -8$$

But what will be 2^{-3} be? This is not defined in a *content-wise manner* (as we have introduced new operations until now), but rather defined in a *structural manner*, in order that the good properties of raising to the power of natural numbers remain fulfilled. As it was valid that $\dfrac{a^n}{a^m} = a^{n-m}$, provided that n and m are natural numbers and that $n > m$, we want this to be valid also in the calculation of the arbitrary integers m and n. Let us see how to realise this with a concrete example:

$$\frac{2^3}{2^6} = \frac{\not{2} \cdot \not{2} \cdot \not{2}}{\not{2} \cdot \not{2} \cdot \not{2} \cdot 2 \cdot 2 \cdot 2} = \frac{1}{2^3}$$

If we want the mentioned rule for division of powers to remain valid, it has to be

$$\frac{2^3}{2^6} = 2^{3-6} = 2^{-3}$$

Thus, we have to define that

$$2^{-3} = \frac{1}{2^3}$$

Generally, raising to the power of numbers by a negative exponent has to be equal to raising to the power of its reciprocal number by the opposite exponent.

Raising to the Power by Integer

For a rational number a and a positive natural number n:

1. $a^n = a \cdot a \cdot a \cdot \ldots \cdot a$ (n times).

2. If $a \neq 0$, then $a^0 = 1$ (0^0 not defined).

3. If $a \neq 0$, then $a^{-n} = \left(\dfrac{1}{a}\right)^n$.

When the exponent also contains a rational number (e.g. $2^{\frac{1}{2}}$) the set of rational numbers becomes deficient. Such powers will be enabled by real numbers, and this will be considered in the following chapter.

Basic Rules of Raising to the Power

For a and b rational numbers, and n and m integers it is valid:

1. $a^n \cdot a^m = a^{n+m}$

2. $a^n : a^m = a^{n-m}$

3. $a^n \cdot b^n = (ab)^n$

4. $a^n : b^n = (a : b)^n$

5. $(a^n)^m = a^{nm}$

6. $0^n = 0 \quad (n > 0), \ 1^n = 1$

7. *For $a > 1$, $a^n < a^m$ precisely when $n < m$*

8. *For $0 < a < 1$, $a^n < a^m$ precisely when $n > m$*

9. *For $a, b > 0$ and $n > 0$, $a^n < b^n$ precisely when $a < b$*

10. *For $a, b > 0$ and $n < 0$, $a^n < b^n$ precisely when $a > b$*

This is very boring to prove since numerous sub-situations need to be proved. We must separately consider the situations when n and m are positive, when n is positive and m is negative, and when both are negative, and so on. When m and n are positive numbers, proving is the same as in natural numbers – it follows from the definition of raising to the power as shortened multiplication. The rest of the cases are reduced to this one. Let us illustrate this with the proof of the first property when n is positive, and m is negative, thus $m = -k$, where k is a positive natural number. Let us calculate for $n \geq k$:

$$a^n \cdot a^m = a^n \cdot a^{-k} = a^n \cdot \left(\frac{1}{a}\right)^k = a^n \cdot \frac{1}{a^k} = a^n : a^k = a^{n-k} = a^{n+m}$$

For $n < k$ the calculation is somewhat different and I leave it to you.

In the case of raising to the power we have a simple rule for transferring factors from numerator to denominator and vice versa.

The Rule of Transferring Factors from Numerator to Denominator and Vice Versa

$$\frac{a^n}{1} = \frac{1}{a^{-n}} \qquad \frac{1}{a^n} = \frac{a^{-n}}{1}$$

We transfer factors from numerator to denominator by changing the sign of their exponents.

If in the description

$$\frac{a^{-2}b^3}{c^{-4}d}$$

we want to write everything in the numerator, we will get

$$a^{-2}b^3c^4d^{-1}$$

If we want to write it all in the denominator, we will get

$$\frac{1}{a^2b^{-3}c^{-4}d}$$

If we want to eliminate the negative exponents, we will get

$$\frac{b^3 c^4}{a^2 d}$$

Note that the rules for powers are rules for multiplication and division of powers, whereas for addition and subtraction there are no rules. Therefore, they can help only in cases of multiplication and division. Also, the rules of multiplication and division of powers require equal bases or equal exponents.

Example 3.10.1.

1. Let us calculate $\dfrac{\left[\left(\frac{1}{4}\right)^{-3} \cdot 2^{-8}\right]^{-2}}{4 \cdot 2}$

2. Let us simplify $\dfrac{(ab)^4}{\left(a^{-2}b^3\right)^{-3}}$

Solution.

1. We will present all the numbers as powers of base 2, and then we will simplify the expression according to the rules for powers. Only when we have exhausted these rules will we be able to calculate the expression directly:

$$\frac{\left[\left(\frac{1}{4}\right)^{-3} \cdot 2^{-8}\right]^{-2}}{4 \cdot 2} = \frac{\left[\left(2^{-2}\right)^{-3} \cdot 2^{-8}\right]^{-2}}{2^2 \cdot 2} = \frac{\left[2^6 \cdot 2^{-8}\right]^{-2}}{2^3} = \frac{\left[2^{-2}\right]^{-2}}{2^3} = \frac{2^4}{2^3} = 2$$

2. In the order in which we would calculate if we knew a and b, we will handle them according to the rules for powers:

$$\frac{(ab)^4}{\left(a^{-2}b^3\right)^{-3}} = \frac{a^4 b^4}{a^6 b^{-9}} = a^{-2}b^{13} = \frac{b^{13}}{a^2}$$

$\square$

3.11   Density of Rational Numbers

The comparison of rational numbers differs from the comparison of integers because of the property of density.

> ## Density of Rational Numbers
>
> Between every two rational numbers there is a rational number (in fact, infinitely many rational numbers).

We have already seen at the beginning of the section that we can find the number between two numbers by examining their numerators and denominators. However, there is also a way of finding the number that is exactly in the middle between the numbers a and b. This number is $\dfrac{a+b}{2}$ and we call it the **arithmetic mean** of numbers a and b.

Let us show that the arithmetic mean is really exactly between these numbers, precisely in the middle. Let $a < b$. Then it is also true that $\dfrac{a}{2} < \dfrac{b}{2}$, and therefore that $\dfrac{a}{2} + \dfrac{a}{2} < \dfrac{a}{2} + \dfrac{b}{2}$. Thus, $a < \dfrac{a+b}{2}$. In the same way it is proved that $\dfrac{a+b}{2} < b$ (try it yourself). We still need to prove that the arithmetic mean is precisely in the middle, in other words that $b - \dfrac{a+b}{2} = \dfrac{a+b}{2} - a$. This is proved by simplifying both sides.

A faster way of determining the number that stands between the given fractions is enabled by **McKay's Theorem**. It says that for the rational numbers $\dfrac{m}{n}$ and $\dfrac{k}{l}$, with m and k being integers and n and l positive natural numbers, between them is the number $\dfrac{m+k}{n+l}$:

$$\frac{m}{n} < \frac{m+k}{n+l} < \frac{k}{l}$$

Thus, the required fraction is obtained simply – we add the numerators for a new numerator, and denominators for a new denominator. The proof of the theorem is given in the exercises.

Example 3.11.1. Let us find the number which is between the rational numbers $\dfrac{4}{15}$ and $\dfrac{7}{13}$ by applying the arithmetic mean, and then McKay's Theorem.

Solution. The arithmetic mean of these numbers is:

$$\frac{\frac{4}{15}+\frac{7}{13}}{2} = \frac{\frac{52+105}{15\cdot13}}{2} = \frac{\frac{157}{15\cdot13}}{2} = \frac{157}{15\cdot13\cdot2} = \frac{157}{390}$$

With McKay's Theorem it is easier to obtain the number which is between the required numbers:

$$\frac{4+7}{15+13} = \frac{11}{28}$$

The origin of McKay's Theorem is interesting. While teaching mathematics, a teacher was applying the arithmetic mean to calculate the number which is between two fractions. The student McKay noticed that this could be done in a simpler way by adding the numerators and denominators and making from the sums a new fraction. The teacher soon confirmed that McKay was right, and hence McKay's Theorem. Stevie, admit it – it wouldn't be bad if a theorem was named after you *Stevie's Theorem*.

You can read more about rational numbers at https://en.wikipedia.org/wiki/Fraction_(mathematics)#Arithmetic_with_fractions

3.12 Problems and Solutions

Measuring of equal parts

1. Reduce:

   (a) $\dfrac{60}{90}$ (b) $\dfrac{2^3 \cdot 3^5 \cdot 5}{2^3 \cdot 3^4 \cdot 5^2}$

2. Simplify:

   (a) $\dfrac{x^2 y^3}{x^2 y^3}$ (b) $\dfrac{a^2 b - a^3 c}{a^3 b - a^4 c}$

3. If you share your sandwich with a friend and both of you get a half, and then there come three hungry friends and you fraternally share your half with them, how much of the sandwich will you have had? How much will all your friends have eaten?

4. Besides her son Gottlob, Gudrun had also stepchildren: Hänsel and Gretel. So as to prevent the neighbours from saying that she was a bad stepmother, she gave her stepchildren an apple each saying: *Here is an apple each for you, and you both give one half to my Gottlob!* Was this fair?

Comparison

5. Put the numbers in order: $-\dfrac{5}{7}, \dfrac{7}{10}, \dfrac{9}{13}, -\dfrac{7}{9}, 1.$

6. By putting the fractions on a common denominator find the rational number between $\dfrac{1}{3}$ and $\dfrac{1}{2}$.

7. Are the following statements true?

   (a) Not every two rational numbers are comparable.

   (b) There exists the smallest positive rational number.

   (c) Between every two numbers there are at least two more numbers.

8. On the basis of the definition of the comparison of rational numbers prove that for every rational number a and b it is true that either $a < b$ or $b < a$ or $a = b$.

9. The Greek philosopher Zeno of Elea (490-430 B.C.) gave several arguments that completion of movement is impossible. One of the arguments is called dichotomy and it says "In order to get from point A to point B you have to first arrive at the halfway point along the path ($\frac{1}{2}$). However, to get there you have to get first to its halfway point ($\frac{1}{4}$). However, in order to get there you have to get first to its halfway point ($\frac{1}{8}$). But to get there you first have to..... Thus, you can never arrive." Although at first glance the argument seems convincing it is not correct since we do arrive after all. Where does it go wrong?

Addition and Subtraction

10. Calculate:

$$\text{(a)} \quad -\frac{5}{36} + \frac{1}{6} \qquad \text{(b)} \quad -\frac{1}{2} - \left(-\frac{2}{3} - \frac{1}{2} \right)$$

11. Simplify and check the solution using the software *SageMath*:

$$\text{(a)} \quad \frac{4a+3b}{5ab} + \frac{2b-3c}{2bc} - \frac{a-2c}{ac} \qquad \text{(b)} \quad \frac{x^2}{xy+y^2} - \frac{x^2+y^2}{xy} + \frac{y^2}{x^2+xy}$$

12. Dick, Mick and Rick love eating cakes. Each of them can eat a birthday cake by himself. It takes Dick two hours to do that, Mick needs three hours and Rick six hours. If they eat a cake together, how much of the cake will they eat in one hour (assuming that they will not eat faster - due to greediness - than when they are alone)?

13. Niven sleeps a quarter of the day, he spends one sixth of the day at school, one third working at a computer and watching TV, a sixth of the day with friends, and one twelfth of the day eating (this Niven seems somehow familiar). During the remaining time he helps with the housework and does his homework. How much of the day does he spend on housework and homework?

14. On the basis of the definition of addition and subtraction of rational numbers, prove that for the rational numbers a, b and c it is true that $a - b = c$ precisely when $c + b = a$.

Multiplication

15. Calculate:

    (a) $\left(3\dfrac{1}{4}\right) \cdot 16$ (b) $\dfrac{5}{2} - \dfrac{3}{5} \cdot \left(-\dfrac{5}{6} - \dfrac{5}{3}\right)$

16. Simplify and check the solution using the software *SageMath*:

    (a) $\dfrac{24x^2}{35ab^2} \cdot \dfrac{75ab}{36xy}$ (b) $\left(ab + b^2\right) \cdot \dfrac{ab}{a+b}$

17. When making coffee for two persons - himself and Juliet - Romeo adds 3 sugar cubes and 4 spoons of coffee to 2 decilitres of water. What amounts should Romeo use to make coffee for 5 persons (Juliet invited 3 of her friends to visit them)?

18. When Romeo made a lemonade for Juliet, he put lemon syrup into a third of the glass and then topped it up with water. Juliet found this too strong, so Romeo poured out a quarter of the glass and refilled it with water. Now the drink was "just right". How much lemon syrup should Romeo pour next time into the glass to make the drink according to Juliet's taste?

19. Romeo has an income of 4000 kuna. An eighth of the income has to go to the repayment of a loan, and he gives a fourth of the remaining part to Juliet. However, during the month Juliet takes another fourth of the remaining part. How much money is left to Romeo for other necessities?

Division

20. If the pirates pour wine from a hundred-litre barrel into a wooden beaker of $\dfrac{2}{5}$ litres, how many times will the beaker have been filled by the time the barrel is emptied?

21. A hen and a half, in a day and a half, lay an egg and a half. How many eggs do 9 hens lay in 9 days?

22. How can you cut, without measuring, from a $\dfrac{2}{3}$ metres long tape a part $\dfrac{1}{2}$ metres long?

23. In his will, a well-known writer of mathematical books left to his children the copyright for his 17 bestsellers (horror stories have always sold well). He left his older son the copyrights to half of his books, he left his daughter the copyrights to one third of the books (a little bit less since she got married against his will), and he left to his younger son (the prodigal one) the copyrights to a ninth of the books. The attorney who was supposed to carry out the will, completely distraught, turned to me for help. How is he going to give the older son a half of 17 bestsellers, and so forth? Recognising the trick the famous writer had played, and finding out with how many zeroes (following the number 1) the attorney was going to pay me, I lent him this book and instructed him on how to perform the division. Having now eighteen bestsellers he did it easily. The oldest son got $\frac{1}{2} \cdot 18 = 9$, the daughter $\frac{1}{3} \cdot 18 = 6$ and the youngest son $\frac{1}{9} \cdot 18 = 2$ bestsellers. Thus, he successfully divided 9+6+2=17 of the writer's bestsellers. Everybody received more than they had expected and the attorney returned my bestseller to me. So, do you give away the trick in the question?

Multiplication and Division

24. Calculate:

    (a) $-\dfrac{10}{9} \cdot \dfrac{-2}{5} \cdot \left(-\dfrac{6}{5}\right)$ (b) $\dfrac{2}{3} : \dfrac{10}{21} \cdot \left(-\dfrac{25}{49}\right)$

25. Simplify and check the solution using the software *SageMath*:

    (a) $\dfrac{a+b}{c} \cdot \dfrac{c+d}{d} : \dfrac{c+d}{c} : \dfrac{a+b}{d}$ (b) $-\dfrac{a}{a+b} \cdot a : \left(-\dfrac{a}{a+b}\right)$

26. Calculate:

    (a) $\dfrac{1\frac{1}{3}}{1\frac{2}{5}}$ (b) $\dfrac{4+\frac{3}{8}}{2}$

27. Simplify and check the solution using the software *SageMath*:

$$\text{(a)} \quad \frac{63xy^4}{14y^3} : 42x^3 y \qquad \text{(b)} \quad \frac{\dfrac{36c^3 d^2}{45ab^3}}{\dfrac{48cd^2}{90a^3 b^2}}$$

28. At a speed of $v_0 = 10$ *m/s*, a car started to accelerate uniformly $a = 4$ *m/s*^{2} and reached a speed of $v = 30$ *m/s*. How much of the path s was travelled if the distance is calculated by the formula $s = \dfrac{v^2 - v_0^2}{2a}$?

29. Are the following statements about rational numbers true?

    (a) $\dfrac{1}{a} = a$

    (b) $a : b = c$ precisely when $a : c = b$

    (c) $\dfrac{\dfrac{a}{b}}{\dfrac{c}{d}} = \dfrac{a \cdot c}{b \cdot d}$

    (d) $\dfrac{-a}{-b} = \dfrac{a}{b}$

30. On the basis of the definition of multiplication and division of rational numbers, prove that $a : b = c$ precisely when $c \cdot b = a$.

31. Prove that $\dfrac{a}{b} : \dfrac{c}{d} = \dfrac{a : c}{b : d}$. In the event that the denominators are the same, this formula provides a simple procedure of division. Apply it in order to calculate $\dfrac{15}{17} : \dfrac{5}{17}$.

32. Prove that $\dfrac{1}{a \cdot (a+1)} = \dfrac{1}{a} - \dfrac{1}{a+1}$.

    Using this result, calculate $\dfrac{1}{1 \cdot 2} + \dfrac{1}{2 \cdot 3} + \dfrac{1}{3 \cdot 4} + \ldots + \dfrac{1}{99 \cdot 100}$.

Equations and Laws of Nature

33. Solve the equations:

    (a) $5(3x-2) = 2[2(3x-1)-(x+4)]+6x$ (b) $\dfrac{x-3}{5} - \dfrac{x+2}{6} = 1$

34. From the formula for the area of a trapezium $P = \dfrac{1}{2}v(a+c)$, express c. If $P = 150\ cm^2$, $v = 10\ cm$, $a = 20\ cm$, how much is c?

35. If $y = \dfrac{u+1}{u+2}$, what is u equal to?

36. From the formula for the pressure of ideal gas $p = \dfrac{n \cdot R \cdot t}{V}$, express the gas constant R.

37. From the formula for the force of the body on a pendulum $F = mg + \dfrac{mv^2}{r}$, express its mass m.

38. Using the substitution method solve the systems:

    (a) $\begin{aligned} y &= 2x+3 \\ 2x + y &= -1 \end{aligned}$ (b) $\begin{aligned} x - y &= 4 \\ 2x - 3y &= 7 \end{aligned}$

39. Using the method of opposite coefficients, solve the systems:

    (a) $\begin{aligned} 10x - 12y &= 5 \\ 5x - 6y &= 3 \end{aligned}$ (b) $\begin{aligned} 6x - 3y &= 1 \\ 8x + 5y &= 7 \end{aligned}$

40. A family of five orders a jumbo pizza. If the father is going to eat 5 times more than any of the three children and twice as much as the mother, into how many parts should the pizza be divided so that everyone gets the share they are entitled to? How much of the pizza does the father eat?

41. A brick weighs 1 kg plus half of a brick more. How much does the whole brick weigh?

42. Wandering through the streets, Donny, Ronny and Jonny came across a bunch of football cards (that a kid had probably lost) of players from the last World Cup. Sonny took a third of the football cards but due to his guilty conscience he returned four. Ronny took a fourth of the remaining football cards. He, too, felt remorse so he had to compromise - he returned 3 football cards. Jonny took half of the remaining football cards. He returned two football cards: not because of a guilty conscience, but because they were dirty with chocolate. If 17 football cards were left on the ground, how many football cards were there at the beginning?

43. (an ancient task, but still topical) A sick Chinese man wished for a watermelon and said to his son "Go to the city and bring me back one watermelon. You just have to keep in mind that on your return you will have to pass four customs officers and each of them will take half of the watermelons that you have at the time." How many watermelons did the son have to buy?

44. Usually, an equation is hidden behind the *magical* games of number guessing. For instance *a magician* tells you to imagine a number. Then he asks you to do several operations with it. For instance, to double it, then to add 2, and to multiply with 3 and subtract 5. Finally, he asks you what the result is of all these operations, says *abracadabra* or *eisequaltomcsquared* and then he "guesses" the number you imagined. How can he do this?

45. Waiting for Juliet, Romeo watched the trains. One train took 12 seconds to pass by. It consisted of 8 wagons and a locomotive, and on every wagon it was written that it was 16 metres long. The locomotive was of the same length as the wagons. What was the train's speed?

46. In a family, every son has the same number of brothers and sisters, and every daughter twice as many brothers than sisters. How many children are there in the family?

47. Grandmother wants to give a certain number of mandarins to the children. If every child were to get 6 mandarins, there would be one mandarin too few, and if every child got 5 mandarins, then there would be 3 mandarins too many. How many mandarins, how many children?

48. In two identical barrels there are different amounts of water. The first barrel contains 40 kg water less than the other one and its whole

weight amounts to $\dfrac{5}{6}$ of the entire weight of the second barrel. If we pour the contents of the second barrel into the first one, then the first barrel will be ten times heavier than the second empty barrel. How much does the empty barrel weigh, and how much water was there in each one at the beginning?

49. Marin is 24 years old. He is now twice as old as Dana was when Marin was as old as Dana is now. How old is Dana?

50. The sum of the digits of a three-digit number is 12. The digit of the tens is half of the sum of the other two digits. If we increased the given number by 396 we would obtain the number written with the same digits but in the reverse order. What is the number?

51. A cyclist had a puncture two thirds of the way along the path. He walked the rest of the path and it took him twice as much time as it would have taken if he had ridden the bike. How much faster was the speed of cycling than the speed of walking?

52. A mountaineer climbed a mountain at a speed of 3 kilometres per hour, and went down at a speed of 5 kilometres per hour. Another one climbed and went down at a constant speed of 4 kilometres per hour. If they started at the same time, which mountaineer came back first?

Percentages

53. Express the proportions as percentages: (a) $\dfrac{1}{4}$, (b) $\dfrac{7}{25}$, (c) $\dfrac{2}{5}$, (d) $\dfrac{19}{20}$.

54. Calculate (a) 15% of 260, (b) 12% of 1750, (c) 30% of 2790, (d) 60% of 95.

55. Express the parts of the quantities in percentages: (a) 5 of 100, (b) 28 of 560, (c) 13 of 52, (d) 64 of 1600.

56. Calculate the whole quantity if its (a) 5% is equal to 5, (b) 13% is equal to 260, (c) 25% is equal to 316, (d) 22% is equal to 550.

57. Mateo claims to his business partner "When you add 22% VAT (Value Added Tax) rate to the price of your product and then put your product at a discount of 22%, you will get the initial price." Is Mateo right? Verify his claim using a product whose initial price is 100 kuna.

58. The price of a shirt is reduced by 25%, and after 30 days it is reduced once more by 25%. If the initial price was 1000 kuna, what is the price after both discounts? What is the total discount rate?

59. The price of one litre of gasoline was 4 kuna. The first price increase was 25% and the second was 50%. What is the final price? What is the total percentage of the price increase?

60. The recommendation is that the best height for a bicycle seat is 105% of the inner length of the cyclist's leg. If the inner length of Juliet's leg is 80 centimetres, to what height must Romeo adjust the seat of her bicycle?

61. An electricity supplier, in coordination with the state institution for the protection of citizens, concluded that a raised rate of 9% on the electricity supply would be too strong an attack on the citizens' standard of living. They agreed that the rise would be in two steps: the first raise rate would be 6% and the second 3%. The electricity supplier accepted the deal with open arms. The company even offered a slower rise, in three steps, with rates raising by 3% at each step. Why did the supplier offer this?

62. After a heated debate, Number 1, the leader of the secret group TNT, agreed to raise their salary by 50% in relation to the previous month's salary. Everyone was happy except Bob Rock, because he understood that they had been tricked once again. How was that possible?

Power

63. Calculate:

    (a) $(-2)^{-2}$ (b) -2^{-2} (c) $\left(\dfrac{4}{3}\right)^{-2}$

64. Calculate:

    (a) $\dfrac{3^{-5}\cdot 9^{-1}}{27^{-2}}$ (b) $\left(\dfrac{4^3\cdot\left(\frac{1}{2}\right)^3}{\left(\frac{1}{8}\right)^{-3}\cdot 2^{-3}}\right)^{-1}$

65. Simplify:

    (a) $\left(\dfrac{2a^{-3}b^0}{ac^{-2}}\right)^{-4}$ (b) $\dfrac{a^2 b^{-3}}{cd^{-4}}\cdot\dfrac{5a^{-3}c^4}{b^{-4}d^5}$

66. Calculate and check the solution using the software *SageMath*:

$$\left(\dfrac{18^{-3}\cdot 3^7}{2^{-5}}\right)^{-2}\cdot\left[\dfrac{\frac{1}{3}\cdot 2^{-1}}{9^{-1}\cdot 2}\right]^{-1}$$

67. Compare:

    (a) $\left(\dfrac{5}{2}\right)^5$ and $\left(\dfrac{5}{2}\right)^3$ (b) $\left(\dfrac{2}{5}\right)^4$ and $\left(\dfrac{4}{5}\right)^4$ (c) $\left(\dfrac{2}{5}\right)^4$ and $\left(\dfrac{4}{5}\right)^3$

68. Are the following statements true?

    (a) $a^{-n} = -a^n$

    (b) $0^{-n} = 0$

    (c) For $0 < a < 1$, $a^n < a^m$ precisely when $n > m$.

    (d) For $a, b > 0$ and $n < 0$, $a^n < b^n$ precisely when $a > b$.

69. Prove that $a^{-n} = \dfrac{1}{a^n}$ for every integer n.

Density

70. Prove Mc Kay's Theorem: for arbitrary numbers a, b and positive numbers c and d, if $\dfrac{a}{b} < \dfrac{c}{d}$ then $\dfrac{a}{b} < \dfrac{a+c}{b+d} < \dfrac{c}{d}$.

Various Problems

71. Calculate on your own or using the software *SageMath*:

    (a) $\left(\dfrac{4}{7}-\dfrac{3}{14}\right)-\left(\dfrac{1}{2}-\dfrac{2}{3}\right)$　　(b) $\dfrac{2}{3}\cdot\left(\dfrac{1}{2}-\dfrac{1}{6}\right)-\dfrac{1}{2}\cdot\left(\dfrac{1}{3}-1\right)$

72. Simplify on your own or using the software *SageMath*:

    (a) $\dfrac{m^2-mn}{m^2+mn}$　　　　　　　　　　(b) $\dfrac{a^2}{(a-1)(a+1)}\cdot\dfrac{a+1}{a}$

    (c) $\left(\dfrac{1}{x}-\dfrac{1}{y}\right)\cdot\dfrac{1}{x-y}$　　　　　(d) $\left(1-\dfrac{a-b}{a+b}\right)\cdot(a+b)$

    (e) $\left(\dfrac{1}{a}-\dfrac{1}{b}\right):\left(\dfrac{1}{a}+\dfrac{1}{b}\right)\cdot\dfrac{b+a}{b-a}$

73. Solve the following equations on your own or using the software *Sage-Math*:

    (a) $x-2(1-x)=3(2+x)-2x$　　(b) $\dfrac{1-(1-x)}{5}+\dfrac{x}{10}=\dfrac{x}{2}-2$

74. Solve the following systems of linear equations on your own or using the software *SageMath*:

    (a) $\begin{aligned}5x-3y&=12\\-2x+3y&=-3\end{aligned}$　　(b) $\begin{aligned}7x-2y&=2\\3x+4y&=30\end{aligned}$　　(c) $\begin{aligned}2x+3y&=6\\3x+4y&=7\end{aligned}$

75. Calculate on your own or using the software SageMath:

    $$\dfrac{\left(\dfrac{1}{2}\right)^{-3}\cdot\left(\dfrac{1}{4}\right)^{2}\cdot\left(\dfrac{1}{8}\right)^{-1}}{(2\cdot4\cdot8)^{-4}}$$

76. Write $\dfrac{a^{-2}b^3}{c^4d^{-2}}$ so that all the members are in the numerator, and then all in the denominator, and then all without the negative exponent.

77. Simplify on your own or using the software *SageMath*:

    $$\left(\dfrac{2r^2s^{-3}}{-4r^{-2}}\right)^{-2}\left(\dfrac{4^{-1}r^2s^{-3}}{2^{-1}r}\right)^{-3}$$

78. Micky and Dicky work at the fast food restaurant *Home-made Food*. Micky makes a sandwich every 5 minutes, and Dicky every 6 minutes. If they work 12 hours a day, having a half-hour break twice, how many sandwiches will they produce in one day?

79. A car consumes in city-driving 14 litres of fuel per 100 km, and during open-road driving 8 litres per 100 km. If after 800 km the fuel consumed amounts to 88 litres, how many kilometres has the car travelled in city-driving?

80. A pool can be filled by one pipe in 10 hours, by a second pipe in 12 hours, and with a third one in 15 hours. If we opened all three pipes at the same time, how long would it take to fill the pool?

81. To buy a scooter Mario has earned over the summer a tenth of the necessary amount; he will get a half from his parents, his grandparents will give him a sixth, and his sister will lend him one fifteenth. How much money does Mario still need in order to become a proud owner of a scooter, if he has earned 3456 kuna this summer?

82. A patient should get 800 mg of antibiotics. The antibiotic concentration in the solution is 1 gram per 5 millilitres. How much of the solution should the patient receive?

83. Daniel gave two female rabbits to his friends as a present. Before that, a half of Daniel's rabbits were of the female gender, and afterwards, only one third. How many rabbits did Daniel have before he gave the present?

84. During the tourist season, two cars started at the same time from Split, but in opposite directions: one towards Zadar, and the other one towards Dubrovnik (Croatian cities on the Adriatic coast). One drove at 10 km/h faster than the other one. If after two hours they were at a distance of 160 km from each other, what were their speeds?

85. George walked to watch the match of his favourite club from home to the stadium at a speed of 3 km/h, and, disappointed, he returned on a bus that drove at a speed of 22 km/h. If George spent two hours travelling, how far is his house from the stadium?

86. For dosing a medicine, doctors sometimes use Cowling's rule, according to which the dose d for a child g years old is $d = \dfrac{D(g+1)}{24}$, where D is the dose of the same medicine for an adult.

(a) If the dose of acetaminophen is 1000 milligrams for an adult, what is the dose for a seven-year old child?

(b) On your own or using the software *SageMath* express from the formula the age of the child using other quantities. Based on the obtained expression, determine the age of a child who takes one fourth of the adult dose.

87. Temperature expressed in Fahrenheit (F) is converted to temperature expressed in Celsius (C) with the following formula: $C = \dfrac{5}{9}(F - 32)$. On your own or using the software *SageMath* express F using C and with the obtained formula express in Fahrenheit the temperature of 100 degrees Celsius.

88. Mr. Ivo has one fifth of his savings in dollars, one eighth in euro, and one tenth in kuna. He invested the rest of the savings in INA's shares (a Croatian oil company in the hands of the Hungarian MOL). What proportion of his savings are invested in shares?

89. Signor Alfonso - in the business world known by his nickname Mafioso - determined, before an important business meeting, to write his will. His wife was pregnant, and since he was strictly traditionally brought up, he considered it important whether it would be a son or a daughter. Therefore, this is how he composed his will. If it were a son he should receive twice as much as the mother. If it were a daughter she should receive half as much as the mother. Unfortunately, Alfonso died while doing business, and his wife gave birth to twins, a son and a daughter. What part of the inheritance should each receive?

90. Mary lent money to her brother at an annual interest rate of 8%. Also, she lent two times as much money to her sister, but at an interest rate two times greater. After one year she received interest of 20 kuna. How much money did she lend to her brother and how much to her sister? How much is the brother's total debt and how much is the sister's total debt?

Solutions

1. (a) $\dfrac{2}{3}$, (b) $\dfrac{3}{5}$

2. (a) 1, (b) $\dfrac{1}{a}$

3. What is left is $\dfrac{1}{8}$. The friends will eat $\dfrac{7}{8}$.

4. Hänsel and Gretel got half an apple each and Gottlob two halves, equalling one apple, and this is not fair.

5. $-\dfrac{7}{9} < -\dfrac{5}{7} < \dfrac{9}{13} < \dfrac{7}{10} < 1$

6. $\dfrac{1}{3} = \dfrac{2}{6} = \dfrac{4}{12}, \quad \dfrac{1}{2} = \dfrac{3}{6} = \dfrac{6}{12}$, so that $\dfrac{5}{12}$ is between these two numbers.

7. (a) No, (b) No,   (c) Yes

8. Let us consider the rational numbers written in fractions with the same denominator $\dfrac{m_1}{n}$ and $\dfrac{m_2}{n}$. Regarding comparativeness of integers, it is either $m_1 < m_2$ or $m_1 = m_2$ or $m_1 > m_2$. However, it is then, per definition of comparison of rational numbers, either $\dfrac{m_1}{n} < \dfrac{m_2}{n}$ or $\dfrac{m_1}{n} = \dfrac{m_2}{n}$ or $\dfrac{m_1}{n} > \dfrac{m_2}{n}$.

9. We know from experience that we will pass along the entire way. We will pass half of the way in half the time, one fourth of the way in one fourth of the time, etc. In order to pass a certain distance we have to pass first half of the distance. However, this does not prevent us from moving, as Zeno argued. His argument is therefore wrong. But the persuasiveness of the argument still shows that there is a certain disharmony between the idea of continuous movement and breaking down the movement into many consecutive events.

10. (a) $\dfrac{1}{36}$, (b) $\dfrac{2}{3}$

11. (a) $\dfrac{26b - 7a}{10ab}$, (b) -1

12. In one hour, the entire cake will be eaten.

13. The activities mentioned take the entire day. Thus, Niven does not do any homework at all, nor does he help with housework. Shame on you, Niven!

14. Let $a = \dfrac{m}{n}$, $b = \dfrac{l}{n}$ and $c = \dfrac{k}{n}$ and let $a - b = c$ i.e. $\dfrac{m}{n} - \dfrac{l}{n} = \dfrac{k}{n}$. Then, according to the definition of subtraction, $\dfrac{m-l}{n} = \dfrac{k}{n}$, so that $m - l = k$. According to the property of inversion of addition and subtraction of integers this means that $k + l = m$. It follows from this that $c + b = \dfrac{k}{n} + \dfrac{l}{n} = \dfrac{k+l}{n} = \dfrac{m}{n} = a$. In the same way you can prove the opposite direction.

15. (a) $\left(3\dfrac{1}{4}\right) \cdot 16 = \dfrac{13}{\cancel{4}} \cdot \cancel{16}^{4} = 13 \cdot 4 = 52$

   (b) $\dfrac{5}{2} - \dfrac{3}{5} \cdot \left(-\dfrac{5}{6} - \dfrac{5}{3}\right) = \dfrac{5}{2} - \dfrac{\cancel{3}}{\cancel{5}} \cdot \dfrac{-\cancel{15}^{3}}{\cancel{6}^{2}} = \dfrac{5}{2} + \dfrac{3}{2} = \dfrac{8}{2} = 4$

16. (a) $\dfrac{30x}{7by}$,   (b) ab^2

17. $7\dfrac{1}{2}$ sugar cubes, 10 spoons of coffee and 5 decilitres of water.

18. Pouring off a fourth of the glass, Romeo poured off also a part of the lemon syrup, precisely a fourth of it. Thus, what remained is $\dfrac{3}{4} \cdot \dfrac{1}{3} = \dfrac{1}{4}$ of the lemon syrup in the glass.

19. Romeo ends up with
$$4000 - \dfrac{1}{8} \cdot 4000 - \dfrac{1}{4} \cdot \dfrac{7}{8} \cdot 4000 - \dfrac{1}{4} \cdot \dfrac{3}{4} \cdot \dfrac{7}{8} \cdot 4000 = \dfrac{3}{4} \cdot \dfrac{3}{4} \cdot \dfrac{7}{8} \cdot 4000 = 1968\dfrac{3}{4} \text{ kuna.}$$
He will have to raise another loan.

20. 250 times

21. 54 eggs

22. $\dfrac{1}{2} : \dfrac{2}{3} = \dfrac{3}{4}$. Thus, $\dfrac{1}{2} = \dfrac{3}{4} \cdot \dfrac{2}{3}$. This means that we need to take three fourths of the tape. We will get one fourth by folding the tape in half and by folding the obtained length once again in half, giving four equal parts. If we tear off one such obtained part, we will be left with three fourths of the tape.

23. The trick lies in the fact that $\frac{1}{2} + \frac{1}{3} + \frac{1}{9} = \frac{17}{18}$ and not 1. However, when we add one more book then the heirs receive $\frac{17}{18} \cdot 18 = 17$ books, precisely as needed, and these we can divide now neatly (and I can have my book back).

24. (a) $-\frac{10}{9} \cdot \frac{-2}{5} \cdot \left(-\frac{6}{5}\right) = -\frac{\cancel{10}^2}{\cancel{9}^3} \cdot \frac{2}{\cancel{5}} \cdot \frac{\cancel{6}^2}{5} = -\frac{8}{15}$,   (b) $-\frac{5}{7}$

25. (a) 1,   (b) a

26. (a) $\frac{20}{21}$,   (b) $\frac{35}{16}$

27. (a) $\frac{3}{28x^2}$,   (b) $\frac{3a^2c^2}{2b}$

28. 100 m

29. (a) No,   (b) Yes,   (c) No,   (d) Yes

30. $\frac{m}{n} : \frac{l}{n} = \frac{k}{n} \rightarrow \frac{m}{n} \cdot \frac{n}{l} = \frac{k}{n} \rightarrow \frac{m}{l} = \frac{k}{n}$. Now we calculate: $\frac{k}{n} \cdot \frac{l}{n} = \frac{m}{l} \cdot \frac{l}{n} = \frac{m}{n}$.
Analogously, the opposite direction is proven as well.

31. $\frac{a}{b} : \frac{c}{d} = \frac{a}{b} \cdot \frac{d}{c} = \frac{ad}{bc} = \frac{a}{c} \cdot \frac{d}{b} = \frac{a}{c} : \frac{b}{d} = \frac{\frac{a}{c}}{\frac{b}{d}} = \frac{a:c}{b:d}$.

Specifically, $\frac{15}{17} : \frac{5}{17} = 15 : 5 = 3$.

32. We calculate: $\frac{1}{a} - \frac{1}{a+1} = \frac{a+1-a}{a(a+1)} = \frac{1}{a(a+1)}$.

$\frac{1}{1 \cdot 2} + \frac{1}{2 \cdot 3} + \frac{1}{3 \cdot 4} + \ldots + \frac{1}{99 \cdot 100} = \frac{1}{1} - \frac{1}{2} + \frac{1}{2} - \frac{1}{3} + \frac{1}{3} - \frac{1}{4} + \ldots + \frac{1}{99} - \frac{1}{100} = 1 - \frac{1}{100} = \frac{99}{100}$

33. (a) $x = 2$,   (b) $x = 58$

34. $c = \frac{2P - av}{v}$, $c = 10$

35. $u = \frac{1 - 2y}{y - 1}$

36. $R = \dfrac{p \cdot V}{n \cdot t}$

37. $m = \dfrac{F \cdot r}{gr + v^2}$

38. (a) $(-1, 1)$,　　(b) $(5, 1)$

39. (a) there is no solution,　　(b) $\left(\dfrac{13}{27}, \dfrac{17}{27}\right)$

40. The pizza needs to be divided into twenty-one parts. Father eats $\dfrac{10}{21}$ of the pizza.

41. 2 kg

42. We follow a number of changes of football cards on the ground:
$$x \xrightarrow{\text{Donny}} \frac{2}{3}x + 4 \xrightarrow{\text{Ronny}} \frac{3}{4}\left(\frac{2}{3}x + 4\right) + 3 \xrightarrow{\text{Jonny}} \frac{1}{2}\left[\frac{3}{4}\left(\frac{2}{3}x + 4\right) + 3\right] + 2$$
Thus the required equation is $\dfrac{1}{2}\left[\dfrac{3}{4}\left(\dfrac{2}{3}x + 4\right) + 3\right] + 2 = 17$. Its solution is $x = 48$.

43. Let x be the unknown number of watermelons. Let us follow the changes in the number of watermelons at the meetings with the four customs officers:
$$x \to \frac{1}{2}x \to \frac{1}{2}\cdot\frac{1}{2}x \to \frac{1}{2}\cdot\frac{1}{2}\cdot\frac{1}{2}x \to \frac{1}{2}\cdot\frac{1}{2}\cdot\frac{1}{2}\cdot\frac{1}{2}x$$
Since he must have one watermelon in the end, this has to be
$$\frac{1}{2}\cdot\frac{1}{2}\cdot\frac{1}{2}\cdot\frac{1}{2}x = 1$$
Hence, we easily determine that he must buy $x = 16$ watermelons.

44. Let us say you imagined the number x. When you perform the mentioned operations consecutively, you get $(2x + 2)\cdot 3 - 5$ which simplified yields $6x + 1$. When you say to the *magician* that you eventually got, for example, 31 he knows that the required number x is the solution of the equation $6x + 1 = 31$. He knows the formula by heart (subtracts from 31 the number 1 and divides by 6) and thus he is able to "guess" that you imagined the number $x = 5$.

45. $v = \dfrac{s}{t} = \dfrac{9 \cdot 16}{12} = 12$ metres per second

46. 7

47. 23 mandarins and 4 children

48. The weight of the barrels is 40 kg. The first barrel contains 160 kg, and the second one 200 kg of water.

49. 18 years of age

50. The number is 246.

51. Let v_b be the cycling speed, and v_p the walking speed. Let s be the total trip and t the bicycle ride time. According to the formula for uniform motion:

$$v_b = \frac{\frac{2}{3}s}{t} = \frac{2s}{3t} \text{ and } v_p = \frac{\frac{1}{3}s}{2t} = \frac{s}{6t}$$

Then the required ratio is

$$\frac{v_b}{v_p} = \frac{\dfrac{2s}{3t}}{\dfrac{s}{6t}} = 4$$

52. On first inspection, it seems that they will arrive at the same time, because the first mountaineer climbed more slowly, but descended faster. However, the climbing time took much longer than the descending time and therefore he moved for a longer time at the speed of 3 kilometres per hour than at the speed of 5 kilometres per hour. Therefore, his average speed is lower than 4 and he will arrive later. We will get the same answer by a precise calculation. Let the time of the first mountaineer be t_1, the second one t_2, the distance to the top d, the speeds of the first mountaineer v_3 and v_5 and the second one v_4:

$$t_1 = \frac{d}{v_3} + \frac{d}{v_5} = \frac{v_3 + v_5}{v_3 \cdot v_5}d, \quad t_2 = \frac{2d}{v_4}$$

It is then

$$t_1 - t_2 = \frac{v_3 + v_5}{v_3 \cdot v_5}d - \frac{2}{v_4}d = \left(\frac{v_3 + v_5}{v_3 \cdot v_5} - \frac{2}{v_4}\right)d = \left(\frac{8}{15} - \frac{2}{4}\right)d > 0$$

and thus the first mountaineer needs more time.

53. (a) 25%, (b) 28%, (c) 40%, (d) 95%

54. (a) 39, (b) 210, (c) 837, (d) 57

55. (a) 5%, (b) 5%, (c) 25%, (d) 4%

56. (a) 100, (b) 2000, (c) 1264, (d) 2500

57. Mateo is wrong: $100 + \dfrac{22}{100} \cdot 100 = 122$, $122 - \dfrac{22}{100} \cdot 122 = 95.6$

58. The final price is $562\dfrac{1}{2}$ kuna. The price of the shirt is reduced by $\dfrac{4375}{100}\%$ $(= 43.75\%)$

59. $4 \xrightarrow{25\%} 4 + \dfrac{25}{100} \cdot 4 = 5 \xrightarrow{50\%} 5 + \dfrac{50}{100} \cdot 5 = 7\dfrac{1}{2}$ kuna. The total rise rate is $87\dfrac{1}{2}\%$. It is a bigger percentage than the sum of particular rise rates (75%). Why?

60. 84 centimetres

61. It seems that the total rise will be the same and that the stepped rise is a humane decision made by the electricity supplier, because the total rise would be realised later and they would lose some money. However, it is not so. Let us take that one household has the expenses of s kuna. After the sudden rise of 9% the expenses would rise by $\dfrac{9}{100}s$. In the two-step rise, the new price would be $s + \dfrac{6}{100}s = \dfrac{106}{100}s$ after the first increase. However, the next increase is an increase in the new (bigger) price! Therefore, the total increase will be bigger than 9%. Let us calculate this. The second increase will be $\dfrac{3}{100} \cdot \dfrac{106}{100}s = \dfrac{318}{10000}s$. So, the total increase is $\dfrac{6}{100}s + \dfrac{318}{10000}s = \dfrac{918}{10000}s = \dfrac{918/100}{1000/100}s = 9.18\%s$ (we have borrowed decimal notation from the next chapter): more than 9%.

62. It is possible. If, in the month before they got nothing, then 50% of zero is zero.

63. (a) $\dfrac{1}{4}$,   (b) $-\dfrac{1}{4}$,   (c) $\dfrac{9}{16}$

64. (a) $\dfrac{1}{3}$,   (b) 8

65. (a) $\dfrac{a^{16}}{16c^8}$,   (b) $\dfrac{5bc^3}{ad}$

66.
$$\left(\frac{18^{-3}\cdot 3^7}{2^{-5}}\right)^{-2}\cdot\left[\frac{\frac{1}{3}\cdot 2^{-1}}{9^{-1}\cdot 2}\right]^{-1}=\left(\frac{(3^2\cdot 2)^{-3}\cdot 3^7}{2^{-5}}\right)^{-2}\cdot\left[\frac{3^{-1}\cdot 2^{-1}}{(3^2)^{-1}\cdot 2}\right]^{-1}=$$

$$=\left(\frac{3^{-6}\cdot 2^{-3}\cdot 3^7}{2^{-5}}\right)^{-2}\cdot\left[\frac{3^{-1}\cdot 2^{-1}}{3^{-2}\cdot 2}\right]^{-1}=[3\cdot 2^2]^{-2}\cdot[3^1\cdot 2^{-2}]^{-1}=$$

$$=3^{-2}\cdot 2^{-4}\cdot 3^{-1}\cdot 2^2=3^{-3}\cdot 2^{-2}=\frac{1}{3^3\cdot 2^2}=\frac{1}{27\cdot 4}=108$$

67. (a) $\left(\dfrac{5}{2}\right)^5 > \left(\dfrac{5}{2}\right)^3$

   (b) $\left(\dfrac{2}{5}\right)^4 < \left(\dfrac{4}{5}\right)^4$

   (c) $\left(\dfrac{2}{5}\right)^4 = \left(\dfrac{2}{5}\right)^3\cdot\dfrac{2}{5}$, $\left(\dfrac{4}{5}\right)^3 = \left(\dfrac{2}{5}\right)^3\cdot 2^3$ $\rightarrow$ $\left(\dfrac{2}{5}\right)^4 < \left(\dfrac{4}{5}\right)^3$

68. (a) No,   (b) No,   (c) Yes,   (d) Yes

69. For $n > 0$ this is the definition of raising to the power for positive exponents. For n = 0 we can easily see that the statement is valid then as well. For n negative we will take that it is $n = -k$, $k > 0$. We need to prove that $[a^{-(-k)} = \dfrac{1}{a^{-k}}$: that it is $a^k = \dfrac{1}{a^{-k}}$. However, per the definition it is $\dfrac{1}{a^{-k}} = \dfrac{1}{\frac{1}{a^k}} = a^k$.

70. Due to the positivity of numbers b and d, from $\dfrac{a}{b} < \dfrac{c}{d}$ it follows that $ad < bc$. We need to show that $\dfrac{a}{b} < \dfrac{a+c}{b+d}$. Due to the positivity of the denominator this is equivalent to the statement $a(b+d) < (a+c)b$, which arranged is $ad < bc$, and we proved this at the beginning. In the same way it is proven also that $\dfrac{a+c}{b+d} < \dfrac{c}{d}$.

71. (a) $\dfrac{11}{21}$,   (b) $\dfrac{5}{9}$

72. (a) $\dfrac{m-n}{m+n}$,   (b) $\dfrac{a}{a-1}$,   (c) $-\dfrac{1}{xy}$,   (d) $2b$,   (e) 1

73. (a) $x = 4$,   (b) $x = 10$

74. (a) $(3, 1)$,   (b) $(2, 6)$,   (c) $(-3, 4)$

75. 2^{26}

76. $\dfrac{a^{-2}b^3}{c^4 d^{-2}} = a^{-2}b^3 c^{-4}d^2 = \dfrac{1}{a^2 b^{-3} c^4 d^{-2}} = \dfrac{b^3 d^2}{a^2 c^4}$

77. $\dfrac{32 s^{15}}{r^{11}}$

78. 242

79. The car travelled a distance of 400 km in city-driving mode.

80. The pool will be filled in 4 hours.

81. Mario needs $\dfrac{1}{6}$ more of the total price, that is $5\,760$ kuna.

82. 4 millilitres

83. For the required number of rabbits k it holds that $\dfrac{1}{2}k - 2 = \dfrac{1}{3}(k - 2)$. Consequently, $k = 8$.

84. 35 km/h and 45 km/h (In the tourist season, traffic is slowed down.)

85. $5\dfrac{7}{25}$ km

86. (a) $\dfrac{1000}{3}$ milligrams,   (b) $g = \dfrac{24d}{D} - 1 = 5$ years

87. $F = \dfrac{9}{5}C + 32$, 100 Celsius = 212 Fahrenheit

88. $\dfrac{23}{40}$

89. Let the part of the inheritance of the son be s, of the daughter k, and of the mother m. The required numbers are the solutions of the following system: $s = 2m$, $m = 2k$, $m + k + s = 1$. This results in $k = \dfrac{1}{7}$, $s = \dfrac{4}{7}$, $m = \dfrac{2}{7}$.

90. Let x be the money lent to her brother. Then $8\% x + 16\% \cdot 2x = 20 \longmapsto x = 50$. So, she lent to her brother 50 kuna and to her sister 100 kuna. The brother's total debt is 54 kuna and the sister's total debt is 116 kuna.

4

Real Numbers
(or how to measure the World)

In this Chapter we will revise our knowledge about real numbers:

- how we imagine them and what their purpose is;

- how we write them down;

- what their relation to rational numbers is;

- how we represent them graphically;

- why and how we compare, add, subtract, multiply and divide them;

- what the basic properties of these operations are;

- why irrational numbers are special;

- how roots and raising to the power with rational exponents are defined.

We will talk about what the number π is, why errors are necessary in calculations and how to deal with them, what measurement units are, what scientific notation is and why it is good, how we measure a plane with numbers and how this allows us to see functions and equations, and what the relation between real numbers and reality is.

You will find out also why the ancient Greeks are outdated, what the most secret Pythagorean secret is, how chaos occurs, why we use metres in Croatia although we had our traditional Croatian elbow, and more.

4.1   Measuring a Line Segment

So far, we have used numbers for various types of measurements. We used natural numbers to measure how many items were contained in finite sets. We used integers to measure oriented quantities. With rational numbers, we measured the parts resulting from the division of a unit into equal parts. However, all these measures feature some restrictions. For instance, it is surely not equally important to the pirates to have four gold medals or four gold bars, nor is half a medal the same as a half of a gold bar. We need a stronger measurement that we will use to measure the unequal parts of unequal items. Now we will describe such *measuring above measurings* which includes all the previous measurements and even more. I will describe it with the example of **measurement of line segments**. Later you will see that the solution to this problem is at the same time also the solution to the general problem of measuring.

Is the line segment from point A to point B (the line segment with end points A and B will be marked $\overline{AB}$) shorter than the line segment $\overline{CD}$? If yes, how much shorter is it? Finally, how long is line segment $\overline{AB}$?

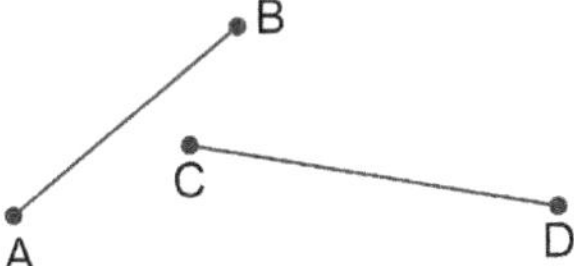

We will be able to answer all the questions if we solve the last one. Since we can put any segment on a line starting from point O (which we will call the **origin**) in the selected direction (which will be called the **positive direction**), we will restrict ourselves to such a situation:

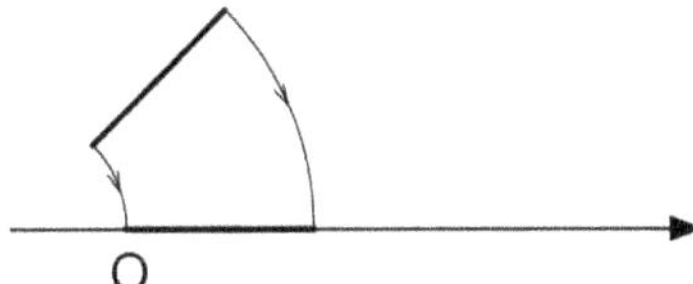

The basic idea is to select one line segment for the unit of measure. It is assigned the value of 1:

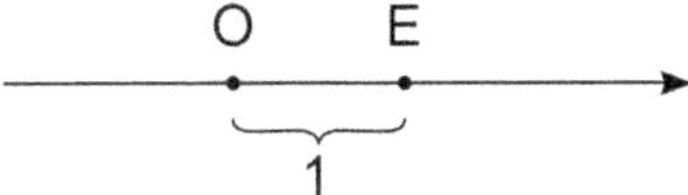

Other segments will be measured on the basis of this one in the following way. The line segment which encompasses twice the unit segment will be assigned the number 2; if it encompasses three times this segment it will be assigned the number 3, and so on. Thus, the natural number n is the measure of the line segment $\overline{OA}$ which encompasses the unit segment exactly n times. We call n the **length** of the line segment $\overline{OA}$ and we mark it $|\overline{OA}|$ or, more simply, $|OA|$. This number is at the same time a measure of the position of point A on the line. It describes it as a point that is n unit segments away from the origin in the selected direction. We call it the **coordinate** of point A on the line:

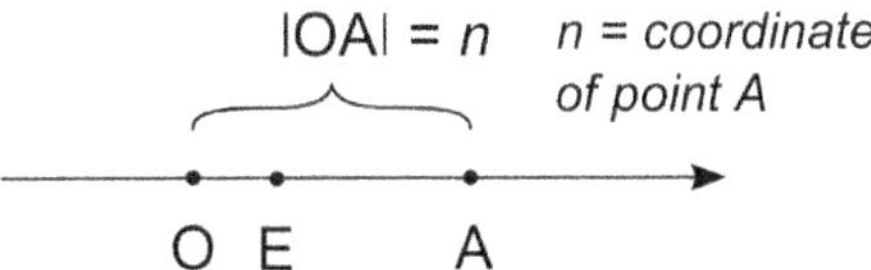

The origin will be assigned the number 0 (since it is precisely this number that measures the distance of a point from itself). The points left of the origin will be assigned negative numbers. If a point has the coordinate –2, it means that it is two unit segments away from the origin, in the direction opposite from the positive direction (this direction will be called the **negative direction**):

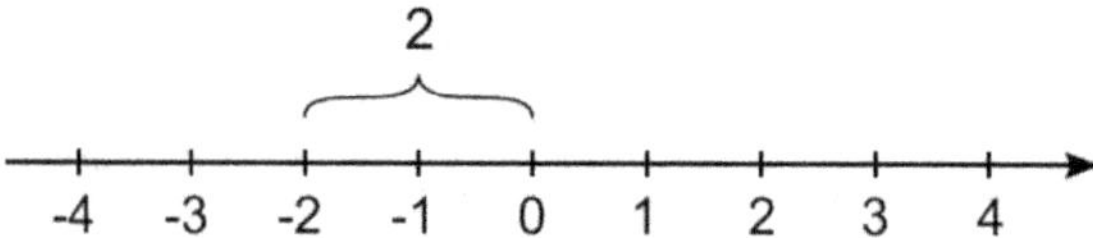

Thus all integers occur as measures of positions of some points on the line.

If we want to measure a certain line segment shorter than the unit one, we have to measure it with parts of the unit segment – halves, thirds, and so on.

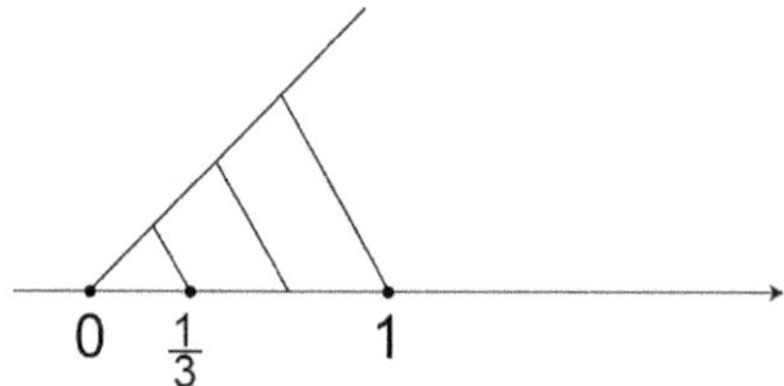

If a third of the unit segment fits into the segment $\overline{OA}$ five times, then we will consider the number $\dfrac{5}{3}$ as its measure:

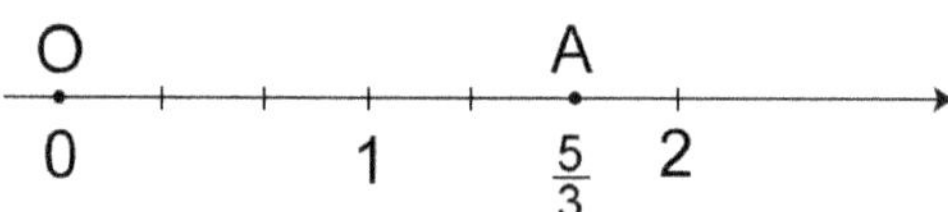

Also, if we go to point B by $\dfrac{3}{2}$ unit segments, but in the negative direction, then its measure is the rational number $-\dfrac{3}{2}$:

In this way, every rational number occurs as a coordinate of a point on the line. However, can we measure the position of every point with a rational number? Is there for every point A an n-th part of the unit which applied m times leads precisely to point A?

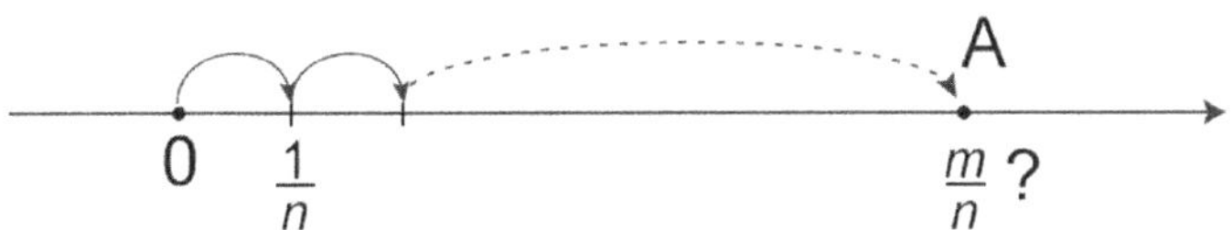

In order to answer this question, we will approach the measurement process in a more systematic way. First, we will determine how many times the unit segment fits into $\overline{OA}$:

We can see that twice is too little, and three times is too much. Therefore, after two units we will continue measuring with the tenth parts of the unit:

This time we can see that $\dfrac{4}{10}$ is too little, and $\dfrac{5}{10}$ is too much. For practical purposes it may be sufficient to say that the line segment length equals approximately $2 + \dfrac{4}{10}$. However, at this point we want to determine the *exact measure* of the segment. Therefore, we will continue measuring with units ten times smaller again: hundredths. We can no longer determine experimentally how many more hundredths fit up to point A, so we will imagine the possible situations. Let us say that after $2 + \dfrac{4}{10}$ there will fit $\dfrac{7}{100}$ more, but that this was too little, whereas $\dfrac{8}{100}$ was too much. In this instance, we will have to continue measuring in thousandths, and even in ten-thousandths, and so on. We can now imagine two possibilities. The first one is that in a certain step, by passing to the units ten times smaller we will measure the remaining part exactly. It may occur that after $2 + \dfrac{4}{10} + \dfrac{7}{100}$ we will need precisely $\dfrac{3}{1000}$ to reach point A. This means that the measure of the segment is the rational number $2 + \dfrac{4}{10} + \dfrac{7}{100} + \dfrac{3}{1000}$. The second possibility is that in no step will we measure exactly the remaining part with a unit ten times smaller. No matter how many times we make the units ten times smaller, point A will always be in between two numbers. Such measurement, in which in every next step we measure with the tenth part of the unit with which we measured in the previous step, is called **decimal measurement**.

4.2 Decimal Notation

Let us analyse in more detail the first possibility. The rational number $2 + \dfrac{4}{10} + \dfrac{7}{100} + \dfrac{3}{1000}$ is more simply written as 2.473. Such notation of a rational number is called **decimal notation** and it is the extension of decimal notation of integers. Each digit shows by its position what it counts. However, now, unlike natural numbers, these do not have to be tens, hundreds and thousands, but may be tenths, hundredths, thousandths, and so on. Thus, in the previous notation, the number 4 shows how many tenths, and 3 how many thousandths there are:

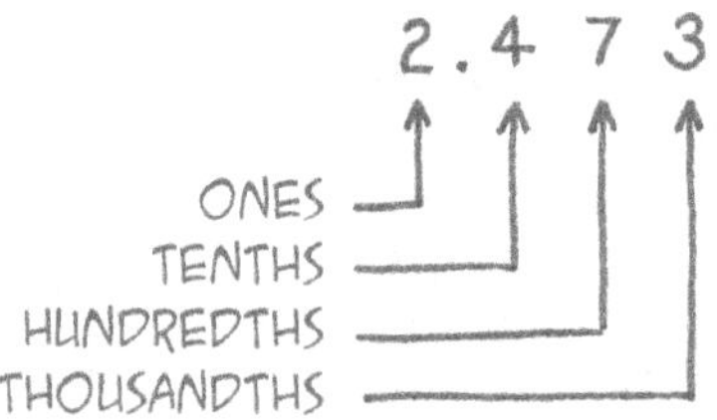

Decimal Notation

In the finite decimal process of measurement there occurs the sum of fractions whose numerators are decimal digits, and in the denominator powers of the number 10. **Decimal notation** is the shortened notation of such a sum:

$$\pm a_0.a_1 a_2 ... a_n = \pm \left(a_0 + \frac{a_1}{10} + \frac{a_2}{10^2} + ... + \frac{a_n}{10^n} \right)$$

where a_0 is a natural number, whereas $a_1, a_2, ..., a_n$ are numbers from 0 to 9.

Thus, decimal notation can be easily transformed into fraction notation. We simply write the values of decimal digits:

$$23.34 = 23 + \frac{3}{10} + \frac{4}{100} = \frac{23 \cdot 100 + 3 \cdot 10 + 4}{100} = \frac{2334}{100}$$

Using this example, we can also observe a faster rule of conversion:

$$23.34 = 2334 \text{ hundreds} = \frac{2334}{100}$$

Conversion of a Decimal Notation into a Fraction

The decimal notation of a number is converted into a fraction so that:

1. we copy the notation without a decimal point into the numerator;

2. we write into the denominator the power of the number 10, which has as many zeroes as there are digits following the decimal point.

This is expressed by the formula:

$$a_0.a_1a_2...a_n = \frac{a_0a_1a_2...a_n}{10^n}$$

In decimal notation, operations with rational numbers are calculated simply. *The procedures are the same as those in calculations with natural numbers, but there are also additional rules for the decimal point.*

Example 4.2.1.

1. Let us compare 0.0357 and 0.0348.

2. Let us add up 3.48 and 34.8.

3. From 0.034 let us subtract 0.019.

4. Let us multiply:

    (a) 4.2 by 100   (b) 3.27 by 4.3

5. Let us divide:

    (a) 4.2 by 100   (b) 43.2 by 0.4

Solution.

1. We search for the first place from the left at which the numbers differ in digits. This is at the position of thousandths. Since the first number has more thousandths, it is bigger (regardless of the fact that the other has more ten-thousandths). Thus,

$$0.0357 > 0.0348$$

2. *Just as in the addition of natural numbers we added in decimal notation – for example, tens with tens – so now we will add also tenths with tenths and possibly convert a part of the results to bigger units. In order to do this properly, we will underwrite the decimal notations so that their decimal points are one below the other. If in adding in some places there are no digits, then we can imagine that these places are occupied by zeroes. Thus,*

$$\begin{array}{r} 3.48 \\ + \ 34.8 \\ \hline 38.28 \end{array}$$

3. As in the case of natural numbers, *we will perform subtraction by moving from right to left subtracting the digits, with the possible loan of a bigger unit:*

$$\begin{array}{r} 0.034 \\ - \ 0.019 \\ \hline 0.015 \end{array}$$

4. (a)
$$4.2 \cdot 100 = \frac{42}{10} \cdot 100 = 42 \cdot 10 = 420$$

   Every digit is assigned the value of a 100 times bigger decimal unit, which corresponds to the shifting of the decimal point 2 places to the right:

$$4.2 \cdot 100 = 420.$$

In general, *multiplication by 10^n means the shifting of the decimal point n places to the right.*

(b)

$$3.27 \cdot 4.3 = \frac{327}{100} \cdot \frac{43}{10} = \frac{327 \cdot 43}{10 \cdot 100}$$

It follows that these need to be multiplied as natural numbers (deleting the decimal point), and the decimal point is added to the result so that it has as many decimal places as both factors together:

$$
\begin{array}{r}
\overset{2}{}\quad\overset{1}{}\\
3.2\!\overset{\frown}{7} \cdot 4.\overset{\frown}{3}\\
\hline
1308\\
981\\
\hline
14.061\\
\underbrace{}\\
3 = 2 + 1
\end{array}
$$

5. (a)

$$4.2 : 100 = \frac{42}{10} \cdot \frac{1}{100} = \frac{42}{1000} = 0.042$$

Every digit obtains the value of a unit 100 times smaller, which corresponds to shifting the decimal point 2 places to the left:

$$4.2 : 100 = 0.042$$

In general, *division by 10^n is the shifting of the decimal point n places to the left.*

(b)

$$4.32 : 0.4 = \frac{4.32}{0.4} = \frac{4.32 \cdot 10}{0.4 \cdot 10} = 43.2 : 4$$

This example shows that *we can simultaneously shift the decimal point of the dividend and the divisor for the same number of places (multiply them or divide them by the same power with base 10), without changing the value of the result. Thus, we can always reduce division in decimal notation to division by natural number, and then the procedure is the same as in the division of natural numbers:*

$$4.32 : 0.4 = 43.2 : 4 = 10.8$$

$$\begin{array}{r} -\,4 \\ \hline 0\,3 \\ -\,0 \\ \hline 3\,2 \end{array}$$

Since it is so easy to calculate decimal notations, why struggle at all with fractions? The problem is that not all rational numbers have a decimal notation. When we converted decimal notation into fractional notation, we obtained in the denominator the powers of the number 10. After cancellation, such denominators when they are factorised into prime factors have the factors 2 and 5. Vice versa, if the fraction in the non-cancellable form has only the factors 2 and 5 in the denominator, then we can expand it to the fraction in which the denominator includes the power of the number 10; thus, we can write it down also in decimal notation.

Example 4.2.2. Let us write down in decimal notation the following fractions:

1. $\dfrac{3}{2}$ 2. $\dfrac{3}{2 \cdot 5^3}$ 3. $\dfrac{5}{3}$

Solution.

1. $\dfrac{3}{2} = \dfrac{3 \cdot 5}{2 \cdot 5} = \dfrac{15}{10} = 1.5$

2. $\dfrac{3}{2 \cdot 5^3} = \dfrac{3 \cdot 2^2}{2^3 \cdot 5^3} = \dfrac{12}{10^3} = \dfrac{12}{1000} = 0.012$

3. The number $\dfrac{5}{3}$ cannot be expanded to the denominator 100, so it has no decimal notation. $\qquad\square$

> **Which Rational Numbers Have a Decimal Notation?**
>
> Not all rational numbers have a decimal notation. The rational number $\frac{m}{n}$, where m and n are non-cancellable natural numbers, has a decimal notation only if the denominator n when broken into prime factors has no other factors except maybe 2 and 5. In this case we can achieve by expansion that the denominator is the power of the number 10, and then the number is easily presented in decimal notation.

You can read more about decimal notation and calculation in this notation at https://en.wikipedia.org/wiki/Decimal.

4.3 Infinite Periodic Decimal Notation

In the previous example we have seen that the rational number $\frac{5}{3}$ has no decimal notation. This means that the decimal process of measuring point A with the coordinate $\frac{5}{3}$ never stops. Although this point has a simple rational measure (a third of the unit segment is applied 5 times and here we are at A), we cannot measure it with decimal parts. In order to see what is happening in the decimal measurement, we have to calculate how many units fit into $\frac{5}{3}$, and then how many tenths, hundredths, and so on. We can discover this using the following process:

$$\frac{5}{3} = 1 + \frac{2}{3} = 1 + \frac{1}{10} \cdot \frac{20}{3} = 1 + \frac{1}{10} \cdot \left(6 + \frac{2}{3}\right) = 1 + \frac{6}{10} + \frac{1}{100} \cdot \frac{20}{3} =$$
$$1 + \frac{6}{10} + \frac{1}{100} \cdot \left(6 + \frac{2}{3}\right) = 1 + \frac{6}{10} + \frac{6}{100} + \frac{1}{1000} \cdot \frac{20}{3} = \dots$$

We can see that in every step, 6 is too small and 7 is too big. However, as the measuring process continues we are getting closer to the number $\frac{5}{3}$. We are creating a sequence of increasing numbers

$$q_0 = 1 < q_1 = 1.6 < q_2 = 1.66 < q_3 = 1.666 < \dots$$

that are getting closer and closer to the number $\frac{5}{3}$:

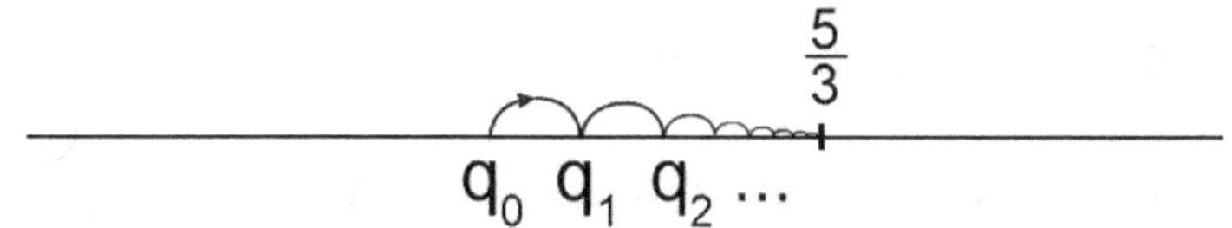

We can also estimate how close we are. The member q_0 is by less than 1 close to $\dfrac{5}{3}$, the member q_1 by less than $\dfrac{1}{10}$, the member q_2 by less than $\dfrac{1}{100}$, and generally, any member q_n is by less than $\dfrac{1}{10^n}$ close to $\dfrac{5}{3}$.

Since the number $\dfrac{5}{3}$ is fully determined by this sequence it is called the **limit of the sequence** $1, 1.6, 1.66, \ldots$. We also say that this sequence **converges** to the number $\dfrac{5}{3}$. We write down

$$\frac{5}{3} = \lim_{n\to\infty} 1.66\ldots 6 \ (n \text{ sixes}), \text{ or more simply}$$

$$\frac{5}{3} = 1.66\ldots, \text{ or even more simply}$$

$$\frac{5}{3} = 1.\bar{6},$$ (the line over the number indicates that the digit 6 is continuously re-
peated in the notation).

This notation describes in a suggestive
manner the members of the sequence and
shows to which number they are getting
closer. Such notation, in which at a certain
place the digits start to be repeated (one
digit or an entire group of digits), which
is emphasised by points or a straight line
above the repeating group of digits, is
called **infinite periodic decimal notation** of the rational number. In order to
be able to distinguish between a plain decimal notation and this notation, the plain
notation is also called **finite decimal notation**.

We will prove now that every rational number has either a finite or infinite periodic decimal notation. This means that the decimal process of measuring the point with a rational coordinate either stops or does not stop, but then the measuring of results by smaller decimal units is repeated periodically.

Let us return to the number $\frac{5}{3}$. The manner in which we have determined its infinite periodic notation is in fact only an extensively written procedure of dividing natural numbers without remainder:

$$
\begin{array}{l}
5 : 3 = 1.666\ldots \\
\underline{-3} \\
20 \\
\underline{-18} \\
20 \\
-18 \\
\underline{} \\
20 \\
\vdots
\end{array}
$$

If we performed this procedure with the fraction $\frac{7}{5}$, the procedure would be completed in one step (the remainder would be 0), since this number has a finite decimal notation:

$$
\begin{array}{l}
7 : 5 = 1.4 \\
\underline{-5} \\
20 \\
\underline{-20} \\
0
\end{array}
$$

But if this procedure was applied to the number $\frac{5}{7}$, it would never stop since the remainder is never 0:

$$5 : 7 = 0.71428517.....$$

$$\frac{-0}{\rightarrow \; \textcircled{5}\,0}$$

$$\begin{array}{r} -49 \\ \hline 10 \\ -7 \\ \hline 30 \\ -28 \\ \hline 20 \\ -14 \\ \hline 60 \\ -56 \\ \hline 40 \\ -35 \\ \hline \rightarrow \textcircled{5}\,0 \\ -49 \\ \vdots \end{array}$$

Since non-zero remainders always have to be smaller than the divisor, at one point a remainder will appear which has already occurred, and the entire process of dividing will be repeated. Here, not only a single digit is repeated, but rather an entire group of digits, so that the infinite periodic notation is as follows:

$$\frac{5}{7} = 0.7142857142857\ldots = 0.\overline{714285}$$

Conversion of a Fractional Notation into a Decimal Notation

The decimal (finite or infinite periodic) notation of a fraction is obtained by the procedure of dividing the numerator and denominator of the fraction.

It follows that *every fraction has a finite or infinite periodic decimal notation*.

Does the reverse hold as well? If, in the infinite process of measuring the position of point A, a group of digits starts to be repeated, is the coordinate of this point a rational number? Let us assume that in the measuring process we obtain the decimal notation

$$2.47353535\ldots$$

and that on the basis of our observations we know that the group of digits '35' will be constantly repeated. Does this array of numbers converge to a rational number or not? If we write this notation extensively in the following way

$$2.47353535\ldots = 2.47 + 0.0035 + 0.000035 + 0.00000035 + \ldots$$
$$= 2.47 + 0.0035 \cdot (1 + 0.01 + 0.0001 + \ldots)$$

we can see that the problem is reduced to the question of whether the array $1.0101\ldots$ tends to a certain rational number. It is easy to ascertain by division that this is the decimal notation of the number $\dfrac{100}{99}$:

$$\frac{100}{99} = 1.010101\ldots$$

Thus, in the infinite periodic measuring process

$$2, 2.4, 2.47, 2.473, 2.4735, 2.47353, 2.473535, 2.4735353, 2.47353535, \ldots$$

we are getting closer and closer to the rational number

$$2.47 + \frac{35}{10000}\frac{100}{99} = 2.47 + \frac{35}{9900}$$

By similar reasoning we can conclude that *every infinite process of decimal measurement in which one group of digits starts to repeat itself describes a point whose coordinate is a rational number.* Following the procedure above, we can calculate such a rational number. In the calculation procedure the problem is reduced to the following formula

$$1.\overline{0\ldots01}(n \text{ decimals}) = \frac{10^n}{10^n - 1}$$

Before arriving at a conclusion, let us consider also a measurement in which the nines are repeated all the time, e.g. $0.9999\ldots$. According to the previous procedure we obtain

$$0.999\ldots = 1$$

We also write

$$\lim_{n\to\infty} 0.\underbrace{999\ldots9}_{n} = 1$$

It seems that in the measurement we made a mistake right at the beginning since we did not observe that the unit segment fits into the measured segment exactly once. We noted that it does not fit exactly once and we went over to measure with tenths, which have to fit then nine times, as well as hundredths, and so on. However, in spite of this mistake, we still managed to measure exactly. It can be proved that such mistakes are correctable only if the nines start to repeat themselves. Every such infinite periodic notation can be simply converted into a finite one:

$$23.47999\ldots = 23.48$$

If we do not allow such correctable mistakes, then every rational number is assigned only one decimal notation, which is either a finite or an infinite periodic notation.

Decimal Notation of Rational Numbers

Every rational number has exactly one finite decimal or infinite periodic decimal notation, if we exclude the notations in which a certain decimal place is followed repeatedly by the digit 9.

Vice versa, every finite decimal or infinite periodic decimal notation is the notation of a rational number.

In *SageMath* we can perform a numerical (decimal) approximation of a rational number (command *n*) and a rational approximation of the number written down in decimal notation (command *Rational*) up to machine precision (sometimes it is an exact conversion). For instance, the commands

```
Rational(0.25),(2/3).n(),Rational(2/3.n()),(Rational(0.25)).n()
```

yield: 1/4, 0.666666666666667, 2/3, 0.250000000000000

4.4 Infinite Non-Periodic Decimal Notation and Real Numbers

We have seen that the process of decimally measuring the position of point A, in other words the segment $\overline{OA}$, can be completed in a number of steps. The segment and the position of the point are measured by the rational number that can be easily read as a result of the measurement process. If, however, the measuring process does not cease in any step, but a group of digits starts to repeat itself instead, then the segment and the position are also measured by the rational number that we can read from the measurement process. But what if the process never ends and if there is no repetition of a group of digits? For instance, measuring point A, we might get the following array:

$$2, 2.4, 2.47, 2.473, 2.4730, 2.47303, 2.473030, 2.4730300, 2.47303003, \ldots$$

If based on some reasoning it follows that we will repeatedly get sequences of zeroes and three, where in every next group there will be one zero more, then the process is infinite and there is no repetition. This sequence determines point A in the same way as an infinite periodic sequence, but since it is non-periodic, the measurement of point A does not yield a rational number! There are two possibilities: either we will accept that certain points have no measure, or we will introduce new numbers with which we will measure such points and segments. The ancient Greeks, for the most part, could not accept new numbers; instead, they concluded that there exist segments that cannot be measured by the unit segment and its parts. They considered only rational numbers as numbers, and these numbers were insufficient for measuring segments. They concluded, therefore, that geometry is stronger than numbers and they turned to geometry. They discovered a lot there, but with this choice they substantially restricted their mathematics. Modern mathematics selected another way which has proved extremely more powerful. The idea about numbers has been expanded so that each segment has its measure.

Real Numbers

Each segment has its measure (**length**). This measure is a **positive real number**. Thus also each point on a straight line on which we select the origin and positive direction has the measure of its position (**coordinate**). Here, different points have different measures. These measures are **real numbers**. We divide them into **positive** (that have the **sign** $+$), **negative** (that have the **sign** $-$) and **zero**. The sign of the coordinate x of the point A says on which side of the origin point A is, and its **amount**, the positive real number $|x|$, how far away it is from the origin.

$$|\,b\,| \qquad |\,a\,|$$

$$b < 0 \qquad 0 \qquad a > 0$$

The set of all real numbers is denoted by $\mathbb{R}$. It includes all the rational numbers. However, there are also real numbers that are not rational. They are called **irrational numbers**. The set of all irrational numbers is denoted by $\mathbb{I}$.

Just as it was valid for rational numbers that they are completely determined by a sequence of rational numbers formed in the process of decimal measurement, so the same holds also for irrational numbers. Thus, it may be said that for all real numbers the following is valid.

Decimal Notation of Real Numbers

For every real number $a \geq 0$ there is a finite or infinite sequence of rational numbers $a_0, a_0.a_1, a_0.a_1a_2, \ldots$ (where $a_1, a_2, \ldots$ are decimal digits, whereas a_0 is a natural number), which is defined by the decimal process of measurement. In addition, $a_0.a_1a_2\ldots a_n$ is nearer to a by less than $\dfrac{1}{10^n}$. The number a is fully determined by this sequence and we designate it as

$$a = a_0.a_1a_2\ldots a_n \ldots (= \lim_{n\to\infty} a_0.a_1a_2\ldots a_n)$$

If a is negative, we can write it down as

$$a = -a_0.a_1a_2\ldots a_n \ldots (= -\lim_{n\to\infty} a_0.a_1a_2\ldots a_n)$$

Thus, for every a there is a sequence of rational numbers $a_0, a_0.a_1, a_0.a_1a_2, \ldots$, (where $a_1, a_2, \ldots$ are decimal digits, whereas a_0 is an integer) such that

$$a = a_0.a_1a_2\ldots a_n \ldots (= \lim_{n\to\infty} a_0.a_1a_2\ldots a_n)$$

Such an array is called a **decimal expansion** and such notation is called **decimal notation** of the number a. For the given number a such notation is unique if we exclude infinite repetition of the digit 9 after a certain place. If this notation is finite or infinite periodic notation, then a is a rational number. If the notation is infinite and non-periodic then a is an irrational number.

Vice versa, every decimal expansion determines a certain real number.

Example 4.4.1. Which of the following numbers are rational, and which are irrational (it is assumed that the digits continue according to the indicated rule):

1. $0.000\ldots$ 2. $-0.12122122212222\ldots$ 3. $3.121221212212122\ldots$

Solution.

1. The notation is periodic, so this is a rational number (zero).

2. There is no repetition because every time there is one more digit 2 in the group, so this is an irrational number.

3. There is constant iteration of the group of digits '12122', and this is a rational number. □

In this description of real numbers there is one weak spot and this is that we compared a real number with the rational members of its decimal expansion, and I still have not described how real numbers are generally compared (this we have described only for rational numbers). This comparison follows naturally from the previous considerations, as well as operations with real numbers, which are the focus of attention in the next Section.

4.5   Arithmetic Operations with Real Numbers

Real numbers are on the one hand directly connected with ideas of space: they are measures of segments and positions also. On the other hand they are determined by decimal expansion: a sequence of rational numbers which occur in the decimal process of measurement. On the basis of these two characteristics, we can define operations with real numbers and determine the properties of these operations. From the geometric aspect, operations with real numbers express certain geometric operations with segments measured by the real numbers, or with points that are their coordinates. Considered via decimal expansion, operations with real numbers are approximated by corresponding operations with their decimal expansions, where approximations are the better the more members are included in the expansion. Precise definitions follow.

Real numbers that are measures of the points which are further to the right on the directed straight line are considered as bigger. This is in accordance with the comparison of rational numbers.

Comparison of Real Numbers

Let a and b be coordinates of points A and B. We say that $\boldsymbol{a < b}$ if point A is left of point B:

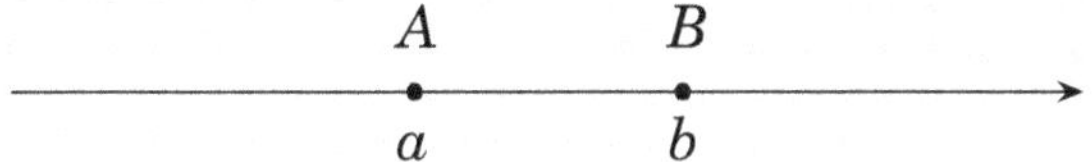

For the positive numbers $a = a_0.a_1a_2\ldots$, and $b = b_0.b_1b_2\ldots$ such that the digit 9 does not appears repeatedly after a certain place, $a < b$ is precisely when, observed from the left, the first place at which the digits differ is such that at this place the digit in the notation of the number a is smaller than the digit in the notation of the number b.

If one of the numbers a and b is negative, the comparison process is the same as for integers.

Applied to rational numbers, this comparison coincides with the previously defined comparison of rational numbers.

Example 4.5.1. Let us compare the following numbers:

1. $\dfrac{2}{3}$ and $0.6767\ldots$ 2. $0,3$ and $0.299\ldots$ 3. $-0.6767\ldots$ and $-0.676776777\ldots$

Solution.

1. In order to compare them, we will convert $\dfrac{2}{3}$ into decimal notation: $\dfrac{2}{3} = 0.6666\ldots.$ The first place from the left where notations differ is the place of hundredths. Since here $\dfrac{2}{3}$ has the digit 6, and the other number has the digit 7, this means that $\dfrac{2}{3}$ is a smaller number: $\dfrac{2}{3} < 0.6767\ldots.$

2. Although the digits at the first decimal place differ, it does not mean that the first number is bigger than the second one, since the rule is valid only when there is no iteration of the nines. And this is precisely what happens here. However, if we write down this number without the nines, $0.2999\ldots = 0.3$, we can see that it is the same number.

3. The comparison of negative numbers is opposite to the comparison of absolute values. Since $0.676767\ldots < 0.676776777\ldots$ it follows that $-0.676767\ldots > -0.676776777\ldots$. $\quad\square$

Addition and subtraction of real numbers is a numerical expression for addition and subtraction of segments.

Addition of Real Numbers

Let a and b be coordinates of points A and B. Then $\boldsymbol{a+b}$ is the coordinate of point C, which we obtain when we move from point A for the segment $\overline{OB}$ to the right if the number b is positive, and to the left if it is negative. If $b = 0$, we stay where we are, at A.

$$| b |, b < 0 \qquad | b |, b > 0$$

$$a + b \qquad\qquad a \qquad\qquad a + b$$

Considered via the decimal expansion of numbers,
for $a = a_0.a_1a_2\ldots$ and $b = b_0.b_1b_2\ldots$,

$$a + b = (a_0.a_1a_2\ldots a_n + b_0.b_1b_2\ldots b_n)\ldots$$

Applied to rational numbers, this addition corresponds to the previously defined addition of rational numbers.

Example 4.5.2. Let us find a sequence which gets closer and closer to the sum $0.9898\ldots + 0.989989998\ldots$.

Solution.

$$0 + 0 = 0, \quad 0.9 + 0.9 = 1.8, \quad 0.98 + 0.98 = 1.96, \quad 0.989 + 0.989 = 1.978,$$
$$0.9898 + 0.9899 = 1.9797, 0.98989 + 0.98998 = 1.97987, \ldots$$

Thus we obtain the sequence:

$$0, 1.8, 1.96, 1.978, 1.9797, 1.97987, \ldots$$

$\quad\square$

There is a problem here. When we add the next members of the sequences, it may result in a change of the previous digits in the result sequence, which is what occurred in the previous example. This means that we did not get a decimal expansion, but rather a different sequence of rational numbers. All the same, we can see that the digits become gradually stabilised and we consider that the sequence has to be nevertheless closer to the required sum $a + b$, which can be also proved.

Opposite Number

Let a be a coordinate of the point A. The opposite number $-a$ is defined as the coordinate of the point B which is at the same the distance from the origin, as well as point A, but on the opposite side.

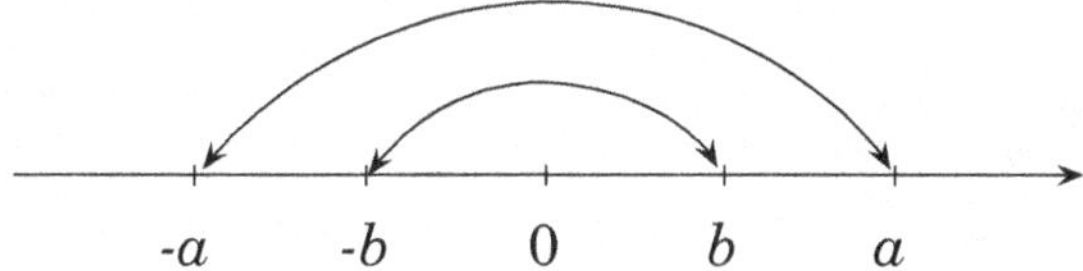

Considered via the decimal expansion of the number:
if $a = a_0.a_1a_2...$, then $-a = -a_0.a_1a_2...$

Here, the following is valid:

1. $|-a| = |a|$ 2. $a + (-a) = 0$ 3. $-(-a) = a$

Applied to rational numbers, this operation corresponds to the previously defined operation of the opposite rational number.

Thus the number opposite to the number $1.232232223...$ is the number $-1.232232223...$, and the number opposite to the number $-4.512511225111222...$ is the number

$$-(-4.512511225111222...) = 4.512511225111222...$$

> ### Subtraction of Real Numbers
>
> Let a and b be coordinates of the points A and B. Then $\boldsymbol{a - b}$ is the coordinate of the point C that is obtained when we move from point A for the segment $\overline{OB}$ to the left if the number b is positive, and to the right if it is negative. If $b = 0$, we remain where we are, at A.
>
> $$|\,b\,|,\, b > 0 \qquad |\,b\,|,\, b < 0$$
>
> $$a - b \qquad\qquad a \qquad\qquad a - b$$
>
> Considered via the decimal expansion of numbers,
> for $a = a_0.a_1a_2\ldots$ and $b = b_0.b_1b_2\ldots,$
>
> $$a - b = (a_0.a_1a_2\ldots a_n - b_0.b_1b_2\ldots b_n)\ldots$$
>
> Subtraction can be described also by using addition:
>
> $$a - b = a + (-b)$$
>
> Applied to rational numbers, this subtraction corresponds to the previously defined subtraction of rational numbers.

Example 4.5.3. Let us find the sequence which is as close as possible to the difference $1.2121\ldots - 2.1212\ldots.$

Solution.

$$1 - 2 = -1,\ 1.2 - 2.1 = -0.9,\ 1.21 - 2.12 = -0.91,\ 1.212 - 2.121 = -0.909,$$
$$1.2121 - 2.1212 = -0.9091,\ 1.21212 - 2.12121 = -0.90909,\ldots$$

Thus we obtain a sequence whose members are closer and closer to the required difference.

$$-1, -0.9, -0.91, -0.909, -0.9091, -0.90909,\ldots$$

$\square$

Geometrically considered, multiplication, for instance by the number 3, corresponds to the extension of the segment, and multiplication by, say, the number $\frac{1}{3}$ corresponds to the contraction of the segment:

The length a can be extended b times in the following manner. We draw two rays from the same apex. We apply length a onto the first ray, whereas we apply the unit length and length b onto the second ray. We connect the points defined by a and 1 and we draw a parallel straight line through the point defined by b. The intersection of this straight line with the first ray defines the length $a \cdot b$:

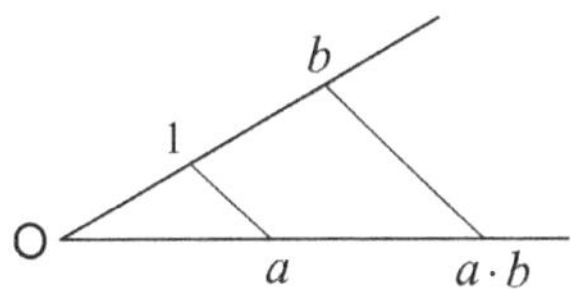

Multiplication of Real Numbers

Let a and b be coordinates of the points A and B.

1. For $b > 0$, $\boldsymbol{a} \cdot \boldsymbol{b}$ is a coordinate of the point C that we obtain by adequate extension/contraction of the segment $\overline{OA}$ presented in the previous figure.

2. For $b < 0$, $\boldsymbol{a} \cdot \boldsymbol{b} = -a \cdot |b|$.

3. For $b = 0$, $\boldsymbol{a} \cdot \boldsymbol{b} = a \cdot 0 = 0$.

Considered via the decimal expansion of numbers,
for $a = a_0.a_1a_2\ldots$ and $b = b_0.b_1b_2\ldots$,

$$a \cdot b = (a_0.a_1a_2\ldots a_n \cdot b_0.b_1b_2\ldots b_n)\ldots$$

Applied to rational numbers, this multiplication corresponds with the previously defined multiplication of rational numbers.

Example 4.5.4. Let us find a sequence which is as close as possible to the product $1.010101\ldots \cdot 1.010010001\ldots$.

Solution. $1 \cdot 1 = 1, 1.0 \cdot 1.0 = 1, 1.01 \cdot 1.01 = 1.0201, 1.010 \cdot 1.010 = 1.0201,$
$1.0101 \cdot 1.0100 = 1.020201, 1.01010 \cdot 1.01001 = 1.020211101,$
$1.010101 \cdot 1.010010 = 1.02021211101, \ldots$

Thus we obtain a sequence whose members get closer and closer to the required product.

$$1, 1, 1.0201, 1.0201, 1.020201, 1.020211101, 1.02021211101, \ldots$$

$\square$

Division is thought of as an opposite operation to multiplication and we can obtain it also geometrically by adequate extension. We draw two rays from the same apex. We apply the length a onto the first ray, and on the second one the unit length and length b. We connect the points defined by a and b and we draw a parallel straight line through the point defined by 1. The intersection of this straight line with the first ray determines the length $\dfrac{a}{b}$:

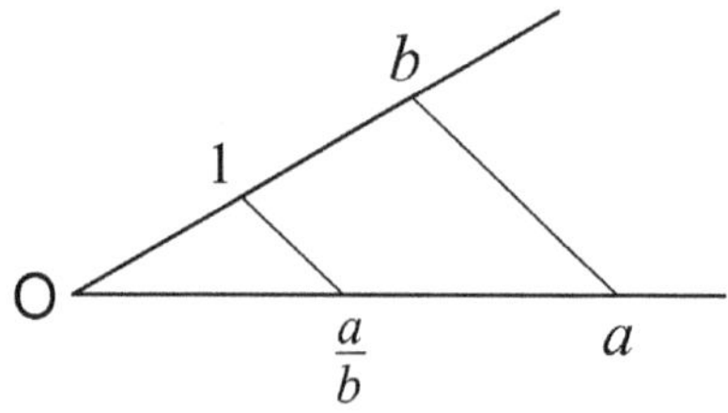

Division of Real Numbers

Let a and b be coordinates of the points A and B.

1. For $b > 0$, $\dfrac{a}{b}$ is the coordinate of the point C that we obtain by adequate extension or contraction of the segment $\overline{OA}$ shown in the previous figure.

2. For $b < 0$, $\dfrac{a}{b} = -\dfrac{a}{|b|}$

3. For $b = 0$, $\dfrac{a}{b}$ is not defined

For $b \neq 0$, the number $\dfrac{1}{b}$ is called the **reciprocal** or **inverse number** of the number b. Division can be described as multiplication by a reciprocal number:

$$\frac{a}{b} = a \cdot \frac{1}{b}$$

Considered via the decimal expansion of numbers,
for $a = a_0.a_1a_2\ldots$ and $b = b_0.b_1b_2\ldots$,

$$\frac{a}{b} = \frac{a_0.a_1a_2\ldots a_n}{b_0.b_1b_2\ldots b_n}\ldots$$

Applied to rational numbers, this division corresponds to the previously defined division of rational numbers.

Example 4.5.5. Let us find the sequence which becomes closer and closer to the number $\dfrac{1.101001000\ldots}{1.101001000\ldots}$.

Solution.

$$\frac{1}{1} = 1, \quad \frac{1.1}{1.1} = 1, \quad \frac{1.10}{1.10} = 1, \ \ldots$$

We divide repeatedly the same numbers so that the result will always be 1. Thus we get the sequence $1, 1, 1,\ldots$ which obviously describes the number 1. $\qquad\qquad\square$

The previous example can be easily generalised: for any $a = a_0.a_1a_2\ldots$ is $\dfrac{a}{a} = 1$. Up to now we have determined this for rational numbers, and now we will for all real numbers. In a similar way, on the basis of the properties of rational numbers we can make conclusions about the properties of real numbers. The other way is geometric. For instance, let us consider the geometric proof that $\dfrac{a}{a} = 1$. We simply apply the definition of division a by b to the division of a by a:

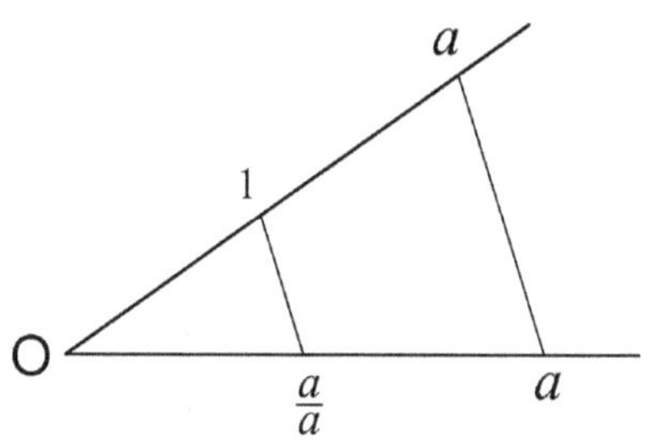

4.6　Basic Properties of Real Numbers

Not only are operations with real numbers an expansion of operations with rational numbers, but they also have equal geometrical meaning. Moreover, operations with irrational numbers can be approximated to operations with rational numbers at will. Therefore, it is no wonder that all the basic properties of rational numbers are valid also for real numbers.

Axioms of Real Numbers

1. $a + b = b + a$ (**commutativity of addition**)

2. $(a + b) + c = a + (b + c)$ (**associativity of addition**)

3. $a + 0 = a$ (**neutrality of zero**)

4. $a - b = c$ precisely when $c + b = a$
   (**inversion of addition and subtraction**)

5. $a \cdot b = b \cdot a$ (**commutativity of multiplication**)

6. $(a \cdot b) \cdot c = a \cdot (b \cdot c)$ (**associativity of multiplication**)

7. $a \cdot 1 = a$ (**neutrality of one**)

8. $a \cdot (b + c) = a \cdot b + a \cdot c$
   (**distributivity of multiplication in relation to addition**)

9. For $b \neq 0$, $a : b = c$ precisely when $c \cdot b = a$
   (**inversion of multiplication and division**)

10. If $a < b$ and $b < c$, then also $a < c$ (**transitivity of comparison**)

11. It is either $a < b$ or $b < a$ or $a = b$ (**comparability**)

12. $a + c < b + c$ precisely when $a < b$
    (**compatibility of comparison and addition**)

13. For $a > 0$, $a \cdot b < a \cdot c$ precisely when $b < c$
    (**compatibility of comparison and multiplication**)

The mentioned axioms are proved either on the basis of fundamental ideas about space or are based on analysis of decimal notations and properties of rational numbers. Using axioms, one can define all the other operations and notions that we considered earlier as basic ones, such as the notion of the opposite number, and prove other properties that we have until now considered as basic ones, for example that $a \cdot 0 = 0$

Example 4.6.1.

1. Let us prove the axiom $a + b = b + a$.

2. Instead of $a : b$ we also write $\dfrac{a}{b}$. Let $b^{-1} = \dfrac{1}{b}$. Let us prove from the axioms that $\dfrac{a}{b} = a \cdot b^{-1}$ (division is multiplication by the reciprocal).

3. Let us prove from the axioms that $a \cdot 0 = 0$.

Solution.

1. Let us consider the case of the positive numbers a and b. Then a is the length of the line segment $\overline{OA}$, and b is the length of the line segment $\overline{OB}$:

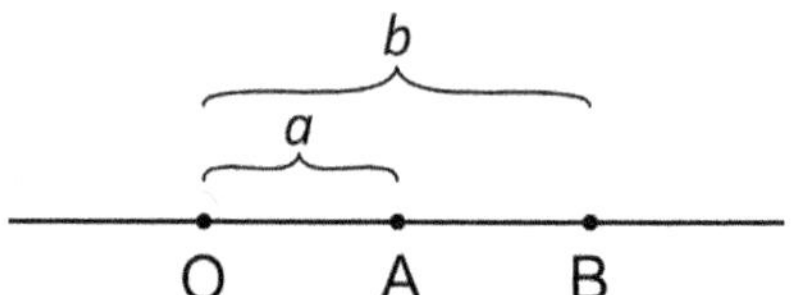

   According to the basic ideas about space, it follows that we will obtain the same segment if we add to the segment $\overline{OA}$ the segment $\overline{OB}$ or if we add to the segment $\overline{OB}$ the segment $\overline{OA}$:

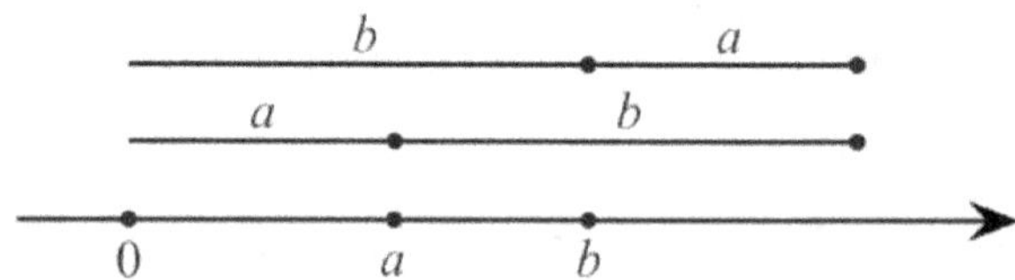

   Commutativity in other cases is reduced to commutativity in the case of the positive numbers a and b.

We could present the proof also on the basis of decimal notation. Let $a = a_0.a_1a_2\ldots$ and $b = b_0.b_1b_2\ldots$. Since due to commutativity of addition of rational numbers

$$a_0 + b_0 = b_0 + a_0, \ a_0.a_1 + b_0.b_1 = b_0.b_1 + a_0.a_1, \ a_0.a_1a_2 + b_0.b_1b_2 = \ldots,$$

these are equal sequences and they tend towards the same number, that is
$a + b = b + a.$

2. From $b^{-1} = \dfrac{1}{b}$ according to axiom 9 it follows that $b^{-1} \cdot b = 1$, and according to commutativity of multiplication also that $b \cdot b^{-1} = 1$. Let $c = a : b$. According to axiom 9, $c \cdot b = a$. If we multiply this equation by b^{-1} we will obtain $(c \cdot b) \cdot b^{-1} = a \cdot b^{-1}$. However, $(c \cdot b) \cdot b^{-1} = c \cdot (b \cdot b^{-1})$ (according to associativity) $= c \cdot 1$ (according to the already proven $b \cdot b^{-1} = 1$.) $= c$ (according to neutrality of the one). Thus, $c = a \cdot b^{-1}$, i.e. $a : b = a \cdot b^{-1}$.

3. According to the neutrality of zero, $a \cdot 0 = a \cdot (0 + 0)$.

   According to distributivity, $a \cdot 0 = a \cdot 0 + a \cdot 0$.

   If we subtract from both sides $a \cdot 0$ we will obtain $a \cdot 0 - a \cdot 0 = a \cdot 0 + a \cdot 0 - a \cdot 0$.

   The parentheses do not have to be written since we could show that subtraction is addition of the opposite and associativity is valid.
   Because for any x, $x - x = 0$ (this can be proved by means of axiom 4), $0 = a \cdot 0 + 0$.

   According to the axiom of the neutrality of zero, $0 = a \cdot 0$. $\qquad\qquad \square$

Let us emphasise one other property that is very important in solving equations.

When the Product Equals Zero

$$a \cdot b = 0 \quad \rightarrow \quad a = 0 \text{ or } b = 0$$

Let $a \cdot b = 0$. We need to prove that if $b \neq 0$ then $a = 0$. Since $b \neq 0$ we can divide by b. According to the axiom of the inversivity of multiplication and division $a \cdot b = 0$ precisely when $a = 0 : b$. However, $0 : b = 0 \cdot b^{-1} = 0$ (we have proved that multiplication by zero yields zero), so that $a = 0$.

A set of objects that contains two distinct objects marked as 0 and 1 and which has four operations highlighted that are marked as +, −, · and :, with valid mentioned axioms 1 to 9, is called a **field**. Such structures often occur in mathematics and allow simple work with equalities. If we also have one relation among the objects which is marked <, with axioms from 10 to 13 being still valid, then we speak about a **linearly ordered field**. Both the world of rational and the world of real numbers are linearly ordered fields. However, the important additional property of real numbers, the one that distinguishes them from rational numbers, should be emphasised once again.

Axiom of Completeness of Real Numbers

14. Every decimal expansion converges to a real number and every real number has a decimal expansion.

Thus, we have obtained the structure of real numbers, which extends the structure of rational numbers. Here, all properties of rational numbers are kept, and moreover measurement has been enabled. A structure which satisfies the 14 mentioned axioms is called a **continuously** or **completely ordered field**.

In *Circle 3* we will show that *the 14 mentioned properties completely specify the world of real numbers*. Then we will also state more precisely what "completely specify" means. Roughly, this means that all the structures (worlds) that satisfy axioms 1 to 14 are not essentially different, the same as there is no essential difference between a wooden chess set and a software simulation of chess on a screen. Whether you play chess with wooden or screen figures makes no difference to the game itself. Because of this, these axioms are called axioms of real numbers. It is from them that all other properties of real numbers are derived (also of rational numbers, integers and natural numbers as subsets of the set of real numbers), and using the notions present in the axioms we define all other notions related to numbers. As you can see from the previous example, completely proving all the other properties would be time-consuming. Therefore, we will often replace them with incomplete proofs or only by assurances that something is. If there is any more serious doubt of the truth of a statement, it can be solved on the basis of the axioms. Also, precise definition of all other notions by means of the notions present in the axioms sometimes takes a long time, so that they will be replaced then by less precise descriptions . If there is any doubt about the meaning of a notion described in this manner, this doubt can be solved by a precise definition.

Axiomatic description of real numbers yields a basis to our conception of numbers. These are simply the basic laws of our conception about numbers. Not being sure of the truth of all the statements about numbers is reduced to insecurity about the truth of these axioms, and insecurity about the meaning of all defined notions about numbers is reduced to insecurity about the meaning of notions that occur in the axioms – zero, one, basic arithmetic operations and comparison. The same holds also for other mathematical worlds.

Furthermore, every statement which we deduce from a set of axioms is valid in every mathematical world that obeys the axioms. For example, the first thirteen axioms for real numbers are also valid for rational numbers. It means that whatever we deduce from them about real numbers, for example that $a \cdot b = 0 \rightarrow a = 0$ or $b = 0$, will also be valid for rational numbers, and for any mathematical world which obeys the axioms. In this way we achieve a great economy of thought and discover similarities between mathematical worlds.

Axiomatic Description

Axiomatic description of any mathematical conception gives this conception maximal precision and uncovers its logical structure. Using the mechanisms of proofs and definitions we discover in a maximally secure manner what is valid in this conception and what is not.

Every statement which we deduce from a set of axioms is valid in every mathematical world which obeys the axioms. In this way we achieve a great economy of thought and discover similarities among mathematical worlds.

In our previous study of numbers we could have been convinced that axiomatic deduction of properties of a mathematical world is more abstract and poorer than the intuitive description of this world, but then on the other hand, more secure and more precise. We have intuitively described natural numbers, integers, rational and real numbers starting from the conception about them as objects for counting and measuring, and only in some places have we used proofs of new properties based on already accepted properties. This is precisely the way mathematics develops:

> **Intuition and Axioms in Mathematics**
>
> Mathematics develops as a combination of intuition and axiomatic deduction. Both components are important and inextricably bound together in the body of mathematics.

In this book (*Circle 1*) as well as in *Circle 2* we will rely more on intuition than on axiomatic deduction, whereas in *Circle 3* and *Circle 4* we will focus more on axiomatic deduction. This is also a natural path in the development of mathematical ideas, from the initial intuition to an axiomatic description that completely specifies the initial intuition. In *Circle 3* we will give, among other things, full axiomatisation of all number structures that we have demonstrated here intuitively. With this, our intuition about numbers acquires a completely precise shape in which there are no confusions about what is and what is not. Proving from the axioms is particularly important in complicated situations and situations about which we have no developed intuition. If we made a comparison with everyday life, *intuition in mathematics is adventure, and axiomatic description is order*.

4.7 Irrational Numbers

The set of real numbers is dense, as well as the set of rational numbers: between any two numbers one can find a rational, as well as an irrational number. The procedure is simple. From the decimal notations of two numbers we will construct the decimal notation of a number between them. It can be a rational as well as an irrational number. The constructions are illustrated in the next example.

Example 4.7.1. Between the numbers $2.344534445344445\ldots$ and $2.345345345\ldots$ let us find one rational and one irrational number.

Solution. Such a rational number is e.g. $2.3446666\ldots$, and an irrational is $2.3446060060006\ldots$. $\qquad\square$

Based on the fact that between every two numbers there is both a rational and an irrational number, it follows that between every two numbers there are infinitely many rational and infinitely many irrational numbers (why?). Moreover, the study of infinite sets has shown that in the exactly definite meaning (which we will leave for *Circle 4*) there are many more irrational numbers than rational ones.

Irrational numbers have occurred until now as possible measures of segments and positions of points. Are there any concrete situations in which irrational numbers occur? For the ancient Greeks they occurred in geometric considerations. Already the simplest right triangle (with two catheti that both have the length 1) "hides" within itself an irrational number - the length of its hypotenuse.

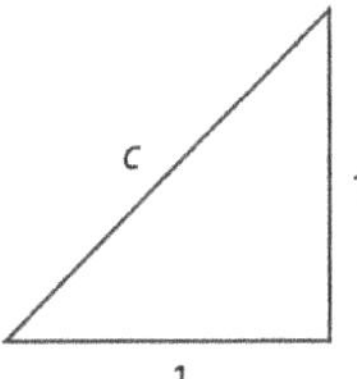

We will prove now that it is indeed so, that whichever part of the unit $\frac{1}{n}$ we take, we will not be able to use it to measure c – that fraction will not fit (the whole number) m times in c. In other words there is no rational number $\frac{m}{n}$ such that $c = \frac{m}{n}$. We will prove this by contradiction. Let us suppose that there were such a rational number $\frac{m}{n}$. Then we could use $\frac{1}{n}$ to be able to measure also x, the number which supplements the unit up to the length of the entire hypotenuse:

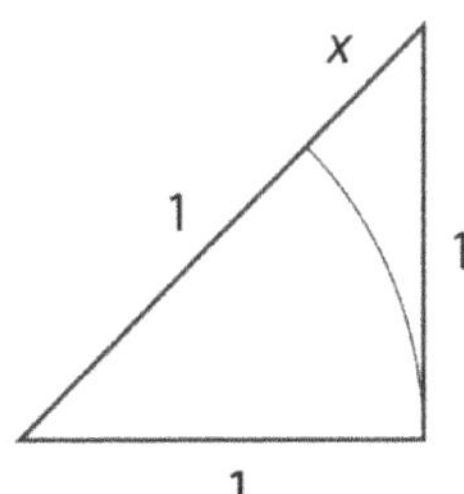

$$x - c - 1 = \frac{m}{n} - 1 = \frac{m - n}{n}$$

It follows that we could measure also $c_1 = 1 - x$ of the hypotenuse of the right triangle that we would obtain by drawing a normal on the hypotenuse at the point that divides the hypotenuse of the initial right triangle into the line segments of lengths 1 and x:

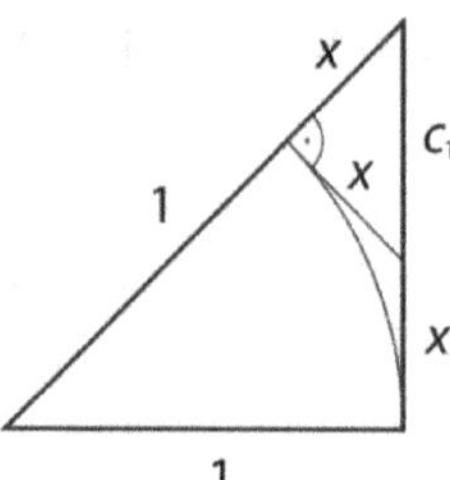

(Think about why three line sections in the figure have the same length x. You don't need to know geometric propositions relevant to this situation – just explore symmetries in the figure.)

Since $c_1 < c$, in order to measure c_1 the number $\dfrac{1}{n}$ should be applied fewer times (m_1 times) than when we measure c (m times): $m_1 < m$. And where is the contradiction? The trick lies in the fact that in the same way in which we found in this bigger right triangle a smaller right triangle, we can find in the smaller right triangle an even smaller one, and so on indefinitely.

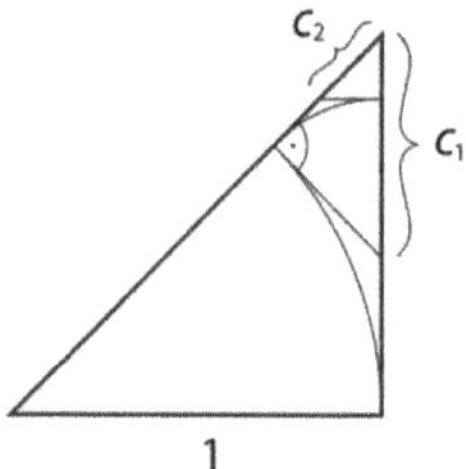

Thus we obtain ever smaller hypotenuses $c > c_1 > c_2 > \ldots$ that we can measure by means of $\dfrac{1}{n}$:

$$\frac{m}{n} > \frac{m_1}{n} > \frac{m_2}{n} > \ldots$$

An ever smaller (natural) number, multiplied by $\dfrac{1}{n}$, fits into them:

$$m > m_1 > m_2 > \ldots$$

By making a sequence of smaller and smaller natural numbers we will obtain a positive natural number m_s which is smaller than 1, and this is a contradiction. Considering the measuring process this would mean that we would get a hypotenuse of length c_s less than $\dfrac{1}{n}$ into which $\dfrac{1}{n}$ fits a (positive) natural number times! Therefore, c (as well as $c_1, c_2, \ldots$) is an irrational number.

This result hit the Greeks' most sensitive spot, in their jewel called **Pythagoras' Theorem**. Pythagoras proved that the sum of the areas of the two squares of the catheti (a and b) equals the area of the square of the hypotenuse:

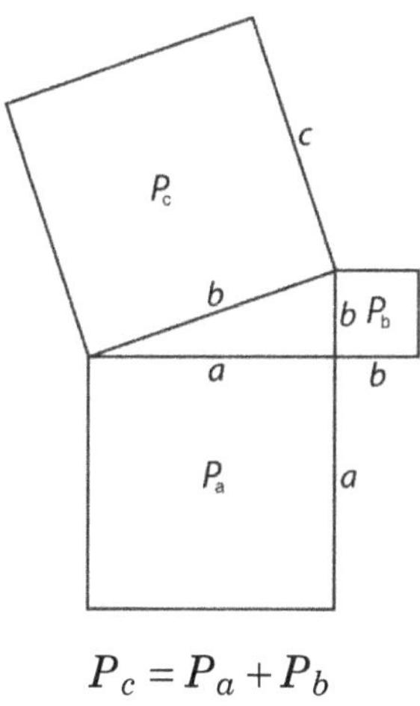

$$P_c = P_a + P_b$$

Since this refers to their squares, it can be expressed in the following way:

$$c^2 = a^2 + b^2$$

The theorem can be proved by geometric comparison of the surfaces into which we break the square of the side $a + b$:

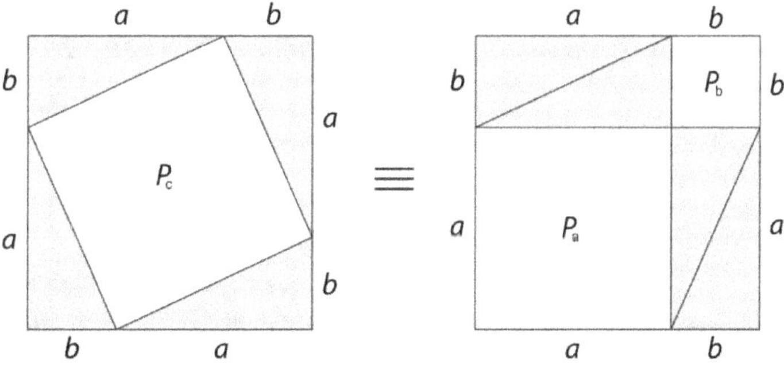

Studying the sounds of a harp, Pythagoras discovered that harmonious tones are produced only by strings whose lengths are in the ratio of natural numbers (for example 2 to 3). This prompted him to think that numbers – particularly rational numbers as ratios of natural numbers – determine the harmony of nature. Moving from the world of sound to the world of vision, he was strengthened in his belief by the theorem he discovered. But then a worm of doubt appeared – some of the simplest segments have no rational measure at all. The story goes that Pythagoras' society was so stricken by this discovery about the imperfection of numbers that they kept it as the darkest secret and that no member of the society was allowed to share it. However, thought cannot be held in captivity. Pythagoras' mistake was not so much in believing that numbers rule the world as in believing that rational numbers rule the world.

Since real numbers are conceived as the measures of segments and positions, the length of the hypotenuse of the right triangle with two equal catheti having the value 1 has its measure which, true enough, is not a rational number. It follows from the Pythagorean theorem that its measure has to be a positive number c which squared yields 2. We designate this irrational number by $\sqrt{2}$:

$$c^2 = 1^2 + 1^2 \;\rightarrow\; c^2 = 2 \;\rightarrow\; c = \sqrt{2}$$

Here is one more simple geometric situation which includes the appearance of irrational number. Just to remind you, a **circle** is a set of all points at an equal distance from one point S that we call the centre of the circle. The distance r of points from the centre is called the *radius of the circle*, and the double radius is called the **diameter of the circle**.

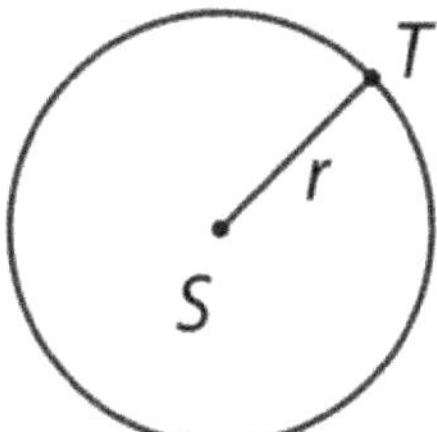

It may be shown geometrically that the ratio of the circumference of a circle and the diameter of the circle is the same number for all circles:

$$\frac{\text{circumference}}{\text{diameter}} = \text{always the same number}$$

This number is written down as π. It has been proved that it is an irrational number. Its approximate value is

$$\pi \simeq 3.14$$

You can read about the irrational history of irrational numbers at `https://en.wikipedia.org/wiki/Irrational_number`.

4.8   Roots and Powers with Rational Exponents

In the same way as for $\sqrt{2}$, we can conclude that there are unique positive numbers which squared yield 3, 4, 5,..., and which we note down with $\sqrt{3}, \sqrt{4}, \sqrt{5}, \ldots$. These can also be understood (according to the Pythagorean theorem) as the lengths of hypotenuses of right triangles that can be obtained by the structure called the **spiral of integer roots** (sounds like the title of some avant-garde song).

Some of these numbers are rational numbers, for instance $\sqrt{4} = 2$. It is easy to discover which numbers these are.

> **When the Root of a Natural Number is a Natural Number**
>
> The root of a natural number n is a natural (rational) number precisely when in its breakdown into prime factors all exponents of prime factors are even numbers. Otherwise, the root is an irrational number.

For example $144 = 2^4 \cdot 3^2$, a $72 = 2^3 \cdot 3^2$, so that $\sqrt{144}$ is a rational number ($= 12$), and $\sqrt{72}$ is not.

We can extend the square root operation from natural numbers to all positive real numbers.

Square Root

For every number $a \geq 0$ there is a unique number $c \geq 0$ such that $c^2 = a$. This number is denoted with $\sqrt{a}$ or $\sqrt[2]{a}$, and it is read the (**square**) **root** of a.

It is easy to see that for $a > 0$ there is at most one positive number c with such a property, since the larger positive number has also a larger square. That such a number does exist, we can prove by decimal measurement. Let us take, say, the number 3. Since the required number c squared yields 3, the unit measurement fits into it once, but not twice, $1^2 < 3 < 2^2$. In the same way we will determine how many tenths fit into it, and so on:

$$1^2 < 3 < 2^2$$

$$1.7^2 < 3 < 1.8^2$$

$$1.73^2 < 3 < 1.74^2 \; 2 < 3 < 2$$

$$1.732^2 < 3 < 1.733^2$$

$$\dots$$

Since the sequence of squares $1^2, 1.7^2, 1.73^2, 1.732^2 \dots$ gets closer and closer to the number 3, so the sequence of numbers $1, 1.7, 1.73, 1.732, \dots$ gets closer and closer to the number (which exists according to the axiom of completeness of real numbers) that squared yields 3 – to the number $\sqrt{3}$.

When a certain operation is defined, we have to be careful that its result is accurately described and that it is not ambiguous. It is therefore important in the definition of the square root for $a > 0$ to define $\sqrt{a}$ as a *positive* number that squared yields a. With the unique positive number c which squared yields a, there is also a unique negative number which squared yields a – this is the opposite number to the number c:

$$(-c)^2 = c^2 = a$$

For instance, there are two numbers that squared yield 4. These are 2 and –2. However, according to the definition of root, $2 = \sqrt{4}$: this allows us to describe also the other number: $-2 = -\sqrt{4}$.

Solving the Equation $x^2 = a$

1. For $a < 0$ there is no solution.

2. For $a = 0$ there is exactly one solution $x = 0$.

3. For $a > 0$ there are two solutions: $x = \sqrt{a}$ and $x = -\sqrt{a}$.

Example 4.8.1. Let us solve the following equations:

1. $x^2 = -100$.   2. $x^2 = 0$ 3. $x^2 = 100$ 4. $x^2 = 10$

Solution.

1. For every x, $x^2 \geq 0$ (why?), so, the equation $x^2 = -100$ has no solution.

2. There is only one number that squared yields zero, and that is 0 (why?). So, $x = 0$.

3. There are two opposite numbers that squared yield 100, and these are 10 and -10.

4. There are two opposite numbers that squared yield 10, and these are $\sqrt{10}$ and $-\sqrt{10}$. $\square$

Higher Roots

If n is an even natural number, then for every number $a \geq 0$ there is a unique number $c \geq 0$ such that $c^n = a$. This number is designated $\sqrt[n]{a}$.

If n is an odd natural number, then for every number a there is a unique number c such that $c^n = a$. This number is designated $\sqrt[n]{a}$.

Example 4.8.2. Let us calculate:

1. $\sqrt[3]{-27}$ 2. $\sqrt[4]{-27}$ 3. $\sqrt[5]{-27}$

Solution.

1. Since $(-3)^3 = -27$ then $\sqrt[3]{-27} = -3$

2. We can find even roots only from non-negative numbers, so that $\sqrt[4]{-27}$ does not exist (nor does $\dfrac{1}{0}$ exist).

3. It is easy to prove, as in studying the square root, that the fifth root of a natural number n is natural (and rational) precisely when in its factorisation into prime factors all exponents are divisible by 5. The number 27 is not such a number, so that $\sqrt[5]{-27}$ has no better name. We can calculate only its decimal expansion which approximates it more and more accurately , as we did with the square root as well:

$$(-2)^5 < -2^7 < (-1.9)^5$$
$$(-1.94)^5 < -2^7 < (-1.93)^5,$$
$$(-1.934)^5 < -2^7 < (-1.933)^5$$

$\cdots\cdots\cdots\cdots$

Thus we calculate the sequence $-1.9, -1.93, -1.933\ldots$ that gets closer and closer to the number $\sqrt[5]{-27}$. $\qquad\square$

The roots also give solutions to certain equations.

Solving the Equation $x^n = a$

For an even number n

1. If $a < 0$ then the equation has no solution.

2. If $a = 0$ then the equation has exactly one solution $x = 0$.

3. If $a > 0$ then the equation has two solutions: $x = \sqrt[n]{a}$ and $x = -\sqrt[n]{a}$.

For an odd number n the equation has exactly one solution $\sqrt[n]{a}$.

Example 4.8.3. Let us solve the following equations:

1. $2x^3 + 2 = 0$ 2. $(2 + x^4) \cdot 8 = 100$

Solution.

1. $2x^3 + 2 = 0 \quad |-2 \quad \rightarrow \quad 2x^3 = -2 \quad |:2$
$\rightarrow x^3 = -1 \quad |\sqrt[3]{\ } \quad \rightarrow \quad x = -1$

2. $(2 + x^4) \cdot 8 = 100 \quad |:8 \quad \rightarrow \quad 2 + x^4 = \dfrac{25}{2} \quad |-2$
$\rightarrow \quad x^4 = \dfrac{21}{2} \quad |\sqrt[4]{\ } \quad \rightarrow \quad x = \pm\sqrt[4]{\dfrac{21}{2}}$ $\qquad\qquad\square$

By means of roots, the powers with rational exponents are defined. Rational numbers have enabled the determining of the powers a^n, where n is an integer. Real numbers enable raising to the power with a rational number. Raising to the power with a rational number is defined in a structural way, so that all rules for powers that were valid also for integers remain valid. This is realised by the following definition.

Raising to the Power by a Rational Number

For $a > 0$ and the natural number n and integer m

$$a^{\frac{m}{n}} = \sqrt[n]{a^m}$$

The idea of raising to the power by irrational number is based on the idea of approximation: the closer the rational number is to the irrational number, the closer is the raising to the power by the rational number to the raising to the power by the irrational number. Since the sequence $1, 1.4, 1.41, \ldots$ gets closer and closer to the number $\sqrt{2}$, it is acceptable to define $2^{\sqrt{2}}$ as a number to which the sequence $2^1, 2^{1.4}, 2^{1.41}, \ldots$ gets closer and closer. The theory of real numbers shows that there is such a unique number.

Raising to the Power by Irrational Number

For $a > 0$ and for the irrational number i such that $i = \lim\limits_{n \to \infty} r_n$, where $r_1, r_2, \ldots, r_n, \ldots$ is a sequence of rational numbers, we define

$$a^i = \lim\limits_{n \to \infty} a^{r_n}$$

Thus defined extension of the notion of power with integer exponent to power with any exponent retains all the good properties of raising to the power.

Basic Properties of Raising to the Power

For a and b positive numbers, and r and s arbitrary real numbers it holds:

1. $a^r \cdot a^s = a^{r+s}$

2. $a^r : a^s = a^{r-s}$

3. $a^r \cdot b^r = (a \cdot b)^r$

4. $a^r : b^r = (a : b)^r$

5. $(a^r)^s = a^{rs}$

6. $1^r = 1$

7. $a^{-r} = \dfrac{1}{a^r}$

8. For $a > 1$, $a^r < a^s$ precisely when $r < s$.

9. For $0 < a < 1$, $a^r < a^s$ precisely when $r > s$.

10. For $a, b, r > 0$, $a^r < b^r$ precisely when $a < b$.

11. For $a, b > 0$ and $r < 0$, $a^r < b^r$ precisely when $a > b$.

We will not prove these properties. I will only illustrate how they can help us in calculating roots. Namely, there are explicit rules for calculating with roots but they are complicated. It is simpler to translate the root notation into a notation of power and calculate according to the rules for powers.

Example 4.8.4. Let us simplify $\sqrt[5]{b^2 \sqrt{b}} : \sqrt[3]{b \sqrt{b}}$

Solution. We will translate the root notation into the notation of powers, simplify according to the rules for powers and return the result back into the root notation.

$$\sqrt[5]{b^2\sqrt{b}} : \sqrt[3]{b\sqrt{b}} = \left(b^2 b^{\frac{1}{2}}\right)^{\frac{1}{5}} : \left(b b^{\frac{1}{2}}\right)^{\frac{1}{3}} = \left(b^{\frac{5}{2}}\right)^{\frac{1}{5}} : \left(b^{\frac{3}{2}}\right)^{\frac{1}{3}} = b^{\frac{1}{2}} : b^{\frac{1}{2}} = 1 \qquad \square$$

In the *SageMath* software, the square root of the number a is calculated by the command sqrt(a) and higher roots using the notation of powers. For instance, for the calculation of $\sqrt[n]{a}$ we use the command a^(1/n).

4.9 Approximate Calculation and Theory of Errors

The problem with calculation with irrational roots, as well as with other irrational numbers, is that they have no standard names, as rational numbers do. For instance, $\sqrt{2}+\sqrt{3}$ has no simpler name. This is simply the sum of positive numbers, one of which squared yields 2 and the other squared yields 3. Such descriptions cannot be calculated as descriptions of rational numbers: rather we can only calculate them by applying the general rules and some special rules for the roots. We can use the decimal notation of irrational numbers only in **approximate calculation**. However, the results are then also approximate. Although in everyday situations we do calculate in this way, more serious situations require **error analysis**. Naturally, we cannot know exactly how much the error is, since then we would obviously know the exact result. What we can calculate is an **error estimate** – which number the error is smaller than, and so between which numbers the correct result is. *Only on this basis can we conclude whether the obtained result is sufficiently good for its desired purpose.* For instance, if we have to walk $\sqrt{2}+\sqrt{3}$ kilometres, we can take that this is approximately

$$\sqrt{2}+\sqrt{3} \simeq 1.414+1.732 = 3.146 \text{ kilometres} = 3146 \text{ metres}$$

Error in calculation is of the size of a metre and this will certainly have no influence on anything, for example, how much food you will take with you on the trip). A common pocket calculator yields sufficiently good approximate results in everyday situations. However, if such calculations decide whether a rocket will land on the Moon or whether it will miss it, then we have to know the error estimate of the approximately obtained result. Only based on this can we conclude whether the approximate result ensures a success-

ful flight of the rocket or not. The analysis of approximate calculation is the subject of the **theory of errors**. This is a very important and not a bit simple part of mathematics and we will only touch on it here.

If we approximated the number a by the number $\overline{a}$, the error would be

$$\Delta = a - \overline{a}$$

If we knew the error we would also know a since

$$a = \overline{a} + \Delta$$

It is more realistic (and sufficient for practical needs) to find the **error estimate**, the number $\varepsilon > 0$ such that $|\Delta| \le \varepsilon$. That for the real value it holds that $\overline{a} - \varepsilon \le a \le \overline{a} + \varepsilon$ we write down more clearly like this:

$$a = \overline{a} \pm \varepsilon$$

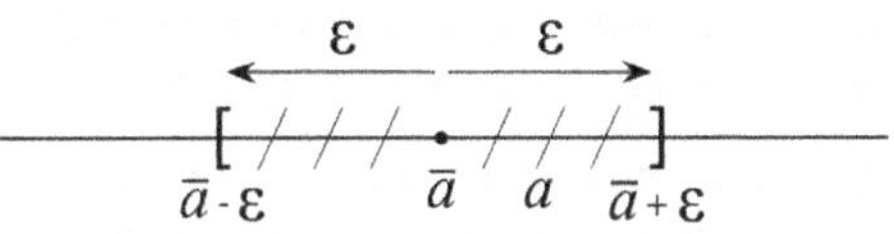

Notation of a Number Using its Approximate Value and the Error Estimate

$$a = \overline{a} \pm \varepsilon$$

For instance, when we write that $\pi = 3.14 \pm 0.01$, it is understood that we will approximate it with the number 3.14 and that the error we make is smaller, regarding absolute value, than 0.01, in other words $3.14 - 0.01 \le \pi \le 3.14 + 0.01$.

Calculating with approximate values, we also obtain approximate results. However, for any operation there is a formula to estimate the error of the obtained result, and so the error estimate can be calculated. In bigger calculations it may happen that the boundaries of the result get completely dissipated, so much so that the result becomes totally non-informative. This is precisely the source of the **theory of chaos**. For instance, weather forecasts are based on experimentally collected data. By the use of the most advanced computers, weather occurrences in the following days are forecast. However, the nature of weather phenomena and the calculation method of these phenomena are such that small errors in experimental data very quickly grow into large errors in calculation results. Therefore, the obtained results are credible only for the following few days, and no further than that.

Exact formulas for error estimates in calculations are very complex. However, there are simple approximate rules (approximation again!) that are satisfactory in practical situations and that we will get to know now.

Often the calculation results **are rounded** at a particular decimal place. This is done by looking at the digit in the decimal place immediately right of the one selected. If that digit is smaller than 5, then the digit in the given decimal place remains as it is, and all the rest of the digits are ignored. If, however, the digit is bigger than or equals 5, then the digit in the selected place is increased by one, and all the following digits are ignored.

Rounding the Number

If the following digit is smaller than 5, the previous digit remains the same, and all the following digits are ignored.

If the following digit is bigger than or equals 5, the previous digit is increased by 1, and all the following digits are ignored.

The rounding error is smaller than the value of the decimal place in which the rounding is done.

In the rules for approximate calculation an important place belongs to **reliable digits** of the number. These are the digits in place values bigger than the error estimate. For instance, if the approximate value of the number 2.0048 is correct up to the second decimal (this means that the error is smaller than a hundredth), the digits 4 and 8 are not reliable since they are within the error estimate, while the other digits are reliable. Therefore, the number is rounded to the reliable digits: $2.0048 \simeq 2.00$. The zeroes following the decimal place are noted in order to point out that these are reliable digits. In approximate calculations only such digits have meaning. When we write that, for instance, $a \simeq 2.37$, we understand that all the mentioned digits are reliable and that the error is smaller than the place value of the digit which is on the extreme right: one hundredth in this example.

Let us note that the number 2.0048 can be written down by adding zeroes at the beginning, for example 02.0048. Zero in the place of hundredths is a reliable digit but not important. Therefore, reliable digits that are not initial (front) zeroes are also called **significant digits**. Let us also notice that the significant digits do not have to be the real digits of the number. They usually are, except maybe the significant digits with the lowest place value. However, if for example we round the number 9.99 to the tenth, we will get the number 10.0, which has completely different digits.

The significant digits let us state simple rules for approximate calculation that liberate us from unnecessary calculations with insignificant digits and yield a rather good error estimate.

Practical Rules of Approximate Calculation

1. **Practical rule of approximate addition and subtraction of several numbers**:

   Among the addends we find the one whose last digit has the biggest place value. It is precisely at this point that we round the result.

2. **Practical rule of approximate multiplication and division of several numbers**:

   Among the factors we find the one that has the fewest significant digits. The result is rounded so that it also has the same number of significant digits.

The rules ensure that the result error in rounding is approximately smaller than the value of the decimal place where the rounding has been performed.

The rules are illustrated in the following example. You can verify their accuracy by marking in every calculation the place of the first insignificant digit with * as a mark of an unknown digit at that place and follow with the calculation procedure which values this affects.

Example 4.9.1. Using the previous rules let us calculate approximately (all the mentioned digits are considered significant):

1. $0.423 + 72.8 + 14.715$ 2. $3.491 \cdot 8.6$ 3. $104.367 + 0.73 \cdot 14.8$

Solution.

1. The last significant digit of the number 72.8 has the biggest place value (tenth) in comparison with the last digits of the other numbers. Therefore, we will round the result to a tenth:

$$0.423 + 72.8 + 14.715 = 87.938 \approx 87.9.$$

2. The number 8.6 has the fewest significant digits – two. Therefore, the result will be rounded to two significant digits:

$$3.491 \cdot 8.6 = 30.0226 \approx 30$$

3. We will first multiply 0.73 and 14.8. The product will be rounded to two significant digits:

$$0.73 \cdot 14.8 = 10.804 \approx 11$$

Now we will add this number to 104.367. We have to round this result to ones:

$$104.367 + 11 = 115.367 \approx 115.$$

$\square$

Approximation of irrational numbers by rational ones is only one of the numerous situations in which we are forced to operate with approximations. As soon as equations become a little bit more complicated, the only efficient way of searching for a solution is to use a numerical procedure (an algorithm which will provide a rational number in the output) to obtain an approximate solution. The same happens also with the calculation of more complex functions, and generally of all the more complex situations in which the solutions are one or several numbers. *Then, out of necessity, we resort to numerical methods that provide us with approximate solutions and an error estimate, because only with the error estimate can the obtained solutions be used credibly.* This part of mathematics is called **numerical mathematics**, and it is a significant tool in solving problems. Since this is rather complicated mathematics, the good news is that there is numerical software into which it is embedded and which can perform it for us.

> **Numerical Mathematics**
> Numerical mathematics = numerical algorithms + error analysis and control

Later we will use numerical software to solve equations and calculate functions (*Circle 2*). We will just briefly describe here the sources of errors that occur while using numerical methods. Errors in the result of a numerical algorithm follow on from errors in both algorithm input data and procedural errors:

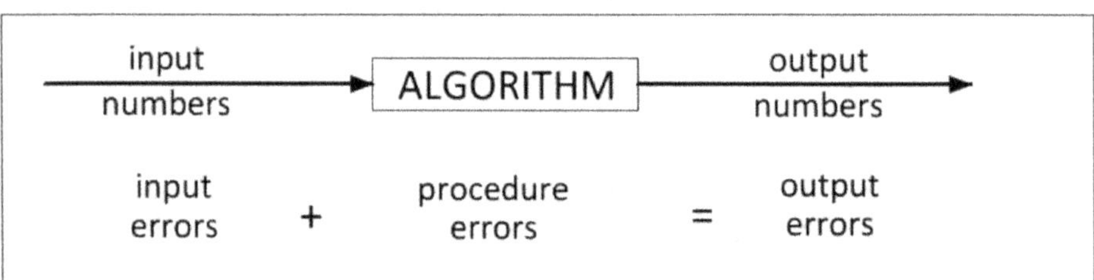

Input errors are of two types. These are experimental errors, and errors of representation. **Experimental errors** result from the procedure of measuring some values that yield an approximate value. For instance, if we use a carpenter's rule to measure a segment, we cannot read a unit smaller than a millimetre, so we have to round the real value to millimetres. **Errors of representation** result from representing real numbers in the computer. In each interval there are infinite numbers and the computer can present them only finitely and we call these numbers **computer numbers**. Thus there is a minimal and a maximal number that can be presented by the computer. If a 64-bit word is used for presenting a number, then these limits are approximately $-2 \cdot 10^{308}$ and $2 \cdot 10^{308}$. Rarely does it occur that this limit is exceeded in common computing. Errors occur because the computer cannot present all the numbers between the mentioned limits. For instance, the smallest positive number that can be presented by a 64-bit computer is approximately $2 \cdot 10^{-308}$. Here, the computer numbers around zero are more densely distributed, and the further we move away from zero the less dense they get:

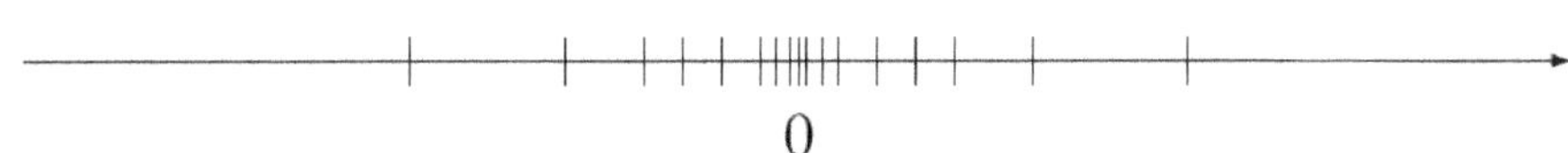

$$0$$

The reason for such a distribution is much greater sensitivity to errors in computing with smaller numbers than with large ones. For instance, if we divide 1 by 0.001 we will obtain 1000, and if we divide by 0.002 we will obtain 500. Thus, a change of input by one thousandth has resulted in a change of 500 in the result.

When we input a number into a computer, it will be represented by the computer number closest to it. The situation is the same as when we compute with numbers on paper. If we agree that we will use in calculation a maximum of, say, 5 decimals, then every number that has more than 5 decimals has to be rounded at the fifth decimal. Thus we will approximate the number to its closest number with 5 decimals. Let it be noted that computers use binary notation. Some numbers with simple decimal notation, such as $\dfrac{1}{10}$, have an infinite binary notation $\left(\dfrac{1}{10} = (0.1)_{10} = (0.0001100110011001\ldots)_2\right)$ and the computer will present them only approximately.

Procedure errors are also of dual type: computing errors and errors in the algorithm itself. **Computing errors** are of the same type as errors of representation. Performing addition and subtraction or multiplication and division, the computer has to represent the obtained result as its closest computer number and this causes error. In individual steps these errors are small, but in some cases, through many steps they can accumulate into a significant error. In order to know exactly what the errors are, an international standard has been established: IEEE 754 (IEEE is the abbreviation for *Institute of Electrical and Electronics Engineers*, a leading international not-for-profit professional organization for technological innovations in the field of electrical engineering). According to this standard, every computer on the market has to comply with certain conditions in the representation of numbers (some of the conditions were mentioned in previous paragraphs) and the size of error of elementary operations. This ensures that on all computers the numerical algorithms will behave in the same manner, so that it is possible to perform a theoretical analysis of computing error regardless of the machine on which the algorithm is being performed.

The **error of a numerical algorithm** is an integral part of its design. Numerical algorithms are usually designed so that the longer they work the better the approximation of the required solution. They stop the moment the required accuracy is achieved. Thus they give an approximate result with an acceptable error. In Circle 2 we will study such algorithms.

Regarding the impact of errors on the algorithm result we distinguish between **stable** and **non-stable** algorithms. In the case of stable algorithms, the errors have an insignificant impact on the result, and in the case of non-stable ones they have a significant impact, so much so that the result is unusable. This does not depend only on the algorithm but also on the problem for which we are searching for a solution. Instances in which small changes cause drastic effects are called chaotic situations. A current example is the traffic in the City of Zagreb (the capital of Croatia). Sometimes it is sufficient that an elderly lady crosses the road at a pedestrian crossing a little too slowly for the roads to get completely congested: some cars do not manage to exit the intersection; therefore others cannot pass through the intersection; thus the adjacent intersection becomes blocked, and so on. The concept that small causes can have large effects is termed **the butterfly effect**.

In the *SageMath* software we have the command n which gives us the numerical approximation of the number to a default *SageMath* precision or to a specified number of significant digits, and the command *round* which rounds the number at a specified decimal space. For instance, the commands

```
pi.n()  # We calculate the number pi to the SageMath
        # default precision.
pi.n(digits = 50) # We calculate the number pi to
                  # 50 significant digits.
round(pi,5) # We round the number pi to the fifth decimal.
```

yield

```
3.14159265358979
3.14159265358979323846264338327950288419716939937510
3.14159
```

You may find more information on numerical mathematics at `https://en.wikipedia.org/wiki/Numerical_analysis`.

4.10 Units of Measurement

Measurement Process

Length measurement represents a sample for measuring other quantities: area, mass, time.... We select one quantity as a unit of measurement and using it in the decimal procedure of measurement, we measure other values.

Naturally, in reality the measurement has to stop once, but in conceptual idealisation the result may be any real number, and not just a rational one. This number says how many units of measurement fit into the measured quantity. For instance:

Measured area $P = 14$ of unit areas

Should we select another value as a unit for measurement, the measurement result will be another number. For instance, let the unit area be the area of square of the side 1 ft long, where ft is the English measure for length, the abbreviation of the English word *foot*. We will denote such a unit ft^2. But why take the English foot when we have our traditional Croatian elbow (we will denote it *Croel*)? Rather, we will take that the unit area is the area of the square of the side one Croel long (we will mark this measure $Croel^2$). Since our Croatian unit Croel is bigger (naturally) than the English foot (a Croel has $\frac{3}{2}$ feet, according to me and my neighbour John), our unit area will be also bigger than the English one. According to the formula for the square area, $Croel^2 = \left(\frac{3}{2}\right)^2 ft^2 = \frac{9}{4} ft^2$. Therefore, the area measurement by Croels will give a smaller number than measuring by feet. The area of $2.5\,Croel^2$ is at the same time also an area of $2.5 \cdot \frac{9}{4} ft^2 = 5.625\,ft^2$. Since neither John nor I agree that the unit measure of the other party should be the official unit measure, we have made the following agreement:

When we assign the number 1 as the measure of a chosen segment, then all other units have their measure: the metre, the Croel and the foot. Let us mark with m the metre measure. Let us assume that by measuring a segment with a metre we obtain that the metre fits into it exactly 5 times. Then its length is $l = 5 \cdot m$. Which number precisely this is depends on which segment we have taken as the unit measure: my elbow (Croel) or John's foot. If we took my elbow (Croel), then m = 2 (since my Croel fits exactly twice into the metre), so that $l = 5 \cdot 2 = 10$. If we took John's foot, since it is $\frac{2}{3}$ of my elbow (Croel), the metre contains $2 \cdot \frac{3}{2} = 3$ John's foot, and in this selection of the unit m = 3 and $l = 5 \cdot 3 = 15$. However, *numbers obtained by the measurement process express only the ratio between the measured value and the value with*

which we have measured. Regardless of which segment we took as the unit, it is always $l = 5 \cdot$ m. The selection of the measurement unit only has an effect on how big the number m is. However, this does not mean anything! Knowing that m is a number assigned to the exactly defined segment (equal in length to the parameter in the Paris Office for Measures), we know exactly how much 5 m is – the length of the segment of a line in which the length of m fits exactly five times. Therefore, we need not highlight any segment as the unit one. The unit length is a "certain" length that is of no interest to us. How many times it fits into the Paris parameter will be called m, in my elbow, it will be called Croel, and in John's foot it will be called ft. What precisely these numbers are, does not matter at all since all other lengths can be expressed by these. There are also formulas for the transformation of one kind of unit into another. Thus,

Units of Measurement

Units of measurement are intentionally non-specified numbers that are measures of exactly specified quantities. Thus, we can calculate with them as with all other numbers, and also other quantities can be measured with them.

Example 4.10.1. Knowing that $m = 2\,\mathrm{Croel} = 3\,\mathrm{ft}$, let us express:

1. the length of 5 m in elbows and feet;

2. the length of 5 Croel in feet and metres;

3. the area of $10\,\mathrm{m}^2$ in square feet (ft^2);

4. the volume of $165\,\mathrm{Croel}^3$ in cubic metres (m^3).

Solution.

1. Applying the conversion formula $m = 2\,\mathrm{Croel}$, we obtain that $5\,m = 5 \cdot 2\,\mathrm{Croel} = 10\,\mathrm{Croel}$. Also, from the conversion formula $m = 3\,\mathrm{ft}$ we obtain that $5\,m = 5 \cdot 3\,\mathrm{ft} = 15\,\mathrm{ft}$.

2. From the conversion formula $2\,\mathrm{Croel} = 3\,\mathrm{ft}$ we will express Croel by means of ft: $\mathrm{Croel} = \dfrac{3}{2}\,\mathrm{ft}$. Now we calculate: $5\,\mathrm{Croel} = 5 \cdot \dfrac{3}{2}\,\mathrm{ft} = \dfrac{15}{2}\,\mathrm{ft}$. Similarly, from the conversion formula $m = 2\,\mathrm{Croel}$ we will express

Croel by means of m: $\text{Croel} = \dfrac{1}{2}\,\text{m}$. Now we calculate again:

$$5\,\text{Croel} = 5 \cdot \frac{1}{2}\,\text{m} = \frac{5}{2}\,\text{m}.$$

3. Applying the conversion formula m = 3 ft, everything is reduced to simple computation:

$$10\,\text{m}^2 = 10 \cdot (3\,\text{ft})^2 = 10 \cdot 9\,\text{ft}^2 = 90\,\text{ft}^2$$

4. From the conversion formula m = 2 Croel we will first express Croel by means of m: $\text{Croel} = \dfrac{1}{2}\,\text{m}$. Now we only have to calculate:

$$16\,\text{Croel}^3 = 16 \cdot \left(\frac{1}{2}\,\text{m}\right)^3 = 16 \cdot \frac{1}{8}\,\text{m}^3 = \frac{16}{8}\,\text{m}^3 = 2\,\text{m}^3$$

Everything stated is valid also for the measuring of other quantities – mass, force, electrical charge, time, energy, and so on.

Conversion of Units of Measurement

We express a certain quantity by the product of the real number and the unit of measurement which we have used for measuring it. The conversion formulas connect units of measurement. They allow us to express a quantity expressed by one unit of measurement by means of other units of measurement.

(The cartoon refers to the so-called conversion of collective ownership into private ownership in the process of transition from socialism to capitalism in Croatia. In simple terms – robbery.)

Example 4.10.2. On a Dalmatian island (bypassed by tourism), apart from kuna the following currencies are used: figs, walnuts, fish, and hammocks. Thus, 50 figs are worth 20 walnuts, 30 walnuts are worth 12 fish, and 100 fish are worth one hammock. How many figs does Signor Domnius have to give for two hammocks (for himself and his wife, Signora Kate)?

Solution. Applying the conversion formulas, we calculate:

$$2\,\text{hammocks} = 2 \cdot 100\,\text{fish} = 200 \cdot \frac{30}{12}\,\text{walnuts} = 500 \cdot \frac{50}{20}\,\text{figs} = 1250\,\text{figs}$$

$\square$

In order to standardise measurements, **an international system of units** was introduced (the SI system). It regulates the basic units, and all others have to be their decimal parts or multiples. Thus, the basic unit for measuring length is a metre, and prefixes determine the larger and smaller units. The numerical meanings of the prefixes are given in the next table.

Prefixes of Measuring Units

$$\text{deka} = 10 \qquad\qquad\qquad \text{deci} = \tfrac{1}{10} = 10^{-1}$$
$$\text{hekto} = 100 = 10^2 \qquad\qquad \text{centi} = \tfrac{1}{100} = 10^{-2}$$
$$\text{kilo} = 1000 = 10^3 \qquad\qquad \text{mili} = \tfrac{1}{1000} = 10^{-3}$$
$$\text{mega} = 1\,000\,000 = 10^6 \qquad \text{micro} = \tfrac{1}{1\,000\,000} = 10^{-6}$$
$$\text{giga} = 1\,000\,000\,000 = 10^9 \qquad \text{nano} = \tfrac{1}{1\,000\,000\,000} = 10^{-9}$$
$$\text{tera} = 1\,000\,000\,000\,000 = 10^{12} \qquad \text{piko} = \tfrac{1}{1\,000\,000\,000\,000} = 10^{-12}$$
$$\dots \qquad\qquad\qquad\qquad \dots$$

Once we have learned the numerical values of the prefixes (which are Greek names for these values), we have at our disposal all the conversion formulas. For instance, in order to see how many centimetres there are in a metre, we just *translate* from Greek:

$$\mathrm{cm} = \frac{1}{100}\,\mathrm{m}$$

Example 4.10.3. Let us convert:

1. $0.25\,\mathrm{km}$ into metres;

2. $0.25\,\mathrm{m}$ into centimetres;

3. $0.025\,\mathrm{km}$ into centimetres;

4. $25\,\mathrm{dm}^3$ into cubic metres;

5. a snail's speed of $1\,\mathrm{cm}$ per minute into kilometres per hour.

Solution.

1. We translate from Greek: $0.25\,\mathrm{km} = 0.25 \cdot 1000\,\mathrm{m} = 250\,\mathrm{m}$.

2. We translate into Greek: $0.25\,\mathrm{m} = 0.25 \cdot 100\,\dfrac{1}{100}\,\mathrm{m} = 25\,\mathrm{cm}$.

3. We convert via the basic unit: $0.025\,\mathrm{km} = 0.025 \cdot 1000\,\mathrm{m} = 25\,\mathrm{m} = 25 \cdot 100\,\mathrm{cm} = 2500\,\mathrm{cm}$.

4. $25\,\mathrm{dm}^3 = 25 \cdot (10^{-1}\,\mathrm{m})^3 = 25 \cdot 10^{-3}\,\mathrm{m}^3 = 0.025\,\mathrm{m}^3$

$$5. \quad 1\frac{\mathrm{cm}}{\mathrm{min}} = \frac{\frac{1}{100}\,\mathrm{m}}{\frac{1}{60}\,\mathrm{h}} = \frac{\frac{1}{100}\cdot\frac{1}{1000}\,\mathrm{km}}{\frac{1}{60}\,\mathrm{h}} = \frac{60\,\mathrm{km}}{100000\,\mathrm{h}} = 0.0006\,\frac{\mathrm{km}}{\mathrm{h}} \qquad \square$$

Although the international system of units enables a standard for communication and simple conversion rules, it should not be understood too bureaucratically. Other units may be used as well. Why shouldn't I compute using my elbows, and my neighbour John using his feet? There are also more serious reasons for the application of other units apart from personal (or national) prestige. Each area of study has its **natural units**. For instance, the sizes of atoms are several $10^{-10}\,\mathrm{m}$ and it is really unnatural to express them with as big a unit as a metre. It would be like shooting a butterfly with a cannon. It was for this reason that the Ångström unit was introduced (as far as I can remember, in honour of a Swedish physicist called Lindström …): $\text{Å} = 10^{-10}\,\mathrm{m}$. This is a unit more appropriate to this field, since the diameters of smaller atoms are several Ångströms. Also, distances between stars are huge and it would be awkward to express them in metres. Here, the natural unit is a light-year. This is the distance travelled by light in one year and it amounts to $\mathrm{ly} = 9.4605284 \cdot 10^{15}\,\mathrm{m}$. Thus, the distance from our planet to the nearest star (if we do not include the Sun) is about 4 light-years, and to the adjacent galaxy Andromeda it is about a million light-years. This means that the image of the galaxy that we currently observe by telescope is actually an image from the past – it shows this galaxy as it was a million years ago. You can read more about measurement units at https://en.wikipedia.org/wiki/Units_of_measurement.

4.11 Scientific Notation

(this sounds scientific)

When computation has to be done with very large and very small numbers, it is good to use so-called scientific notation (thus called since it is usually used in science). A number is presented in the form of the product of numbers between 1 and 9, expressed by a decimal notation, and the power of the number 10. An example is the notation $2.4 \cdot 10^{-5}$. By applying the rule that the multiplication by the powers of the number 10 corresponds to shifting of the decimal point, every decimal notation can easily be presented in scientific notation.

Example 4.11.1. Let us present in scientific notation the numbers:

1. $2\,300\,000$

2. $0.000\,002\,3$

Solution.

1. Shifting of the decimal point has to be compensated for – by multiplying the number with the appropriate power of the number 10: $2\,300\,000 = 2\,300\,000.0 = 2.3 \cdot 10^6$.

2. Shifting of the decimal point to the right means an increase in the number and it has to be compensated for – by multiplying by the power of the number 10 with a negative exponent: $0.000\,002\,3 = 2.3 \cdot 10^{-6}$. $\square$

Scientific Notation of a Number

Scientific notation of a number is notation in the form $a \cdot 10^n$, where $1 \le a < 10$, a is presented in decimal notation, and n is an integer. The conversion of decimal notation into scientific notation is given by the following formulas:

$$a_n a_{n-1} \ldots a_0 . a_{-1} \ldots = a_n . a_{n-1} \ldots \cdot 10^n$$

$$0.00 \ldots a_{-n} a_{-n-1} \ldots = a_{-n} . a_{-n-1} \ldots \cdot 10^{-n}$$

Regarding simpler computations, numbers do not have to be presented precisely in scientific notation. It is sufficient to present them as products of a number close to one (for example in the range from hundredths to thousands) and the power of the base 10. This notation not only simplifies computation but also enables us also to make a rough estimate of the result in advance.

Example 4.11.2.

1. Let us calculate using scientific notation:

$$\frac{0.001 \cdot 5000^2 \cdot 64}{(0.02)^4 \cdot 2\,000\,000\,000}$$

2. During one year the Earth rotates once around the Sun. If the distance between the Earth and the Sun is approximately $150\,000\,000\,\mathrm{km}$ and if the Earth moves approximately along a circle, what is its speed of rotation around the Sun? Let us express the result in kilometres per second.

Solution.

1. $\dfrac{10^{-3} \cdot \left(5 \cdot 10^3\right)^2 \cdot 64}{\left(2 \cdot 10^{-2}\right)^4 \cdot 2 \cdot 10^9} = \dfrac{25 \cdot 64 \cdot 10^{-3} \cdot 10^6}{16 \cdot 2 \cdot 10^{-8} \cdot 10^9} = 50 \cdot 10^2 = 5000$

2. The formula for uniform motion $s = vt$ will help us connect the unknown speed v with the known values $t = 1$ year and the radius of the Earth's path $r = 150\,000\,000\,\mathrm{km}$. Since the Earth moves approximately along a circle, the travelled path s in time $t = 1$ year is precisely the circumference of the circle: $s = 2r\pi$ (according to the formula for the circumference of a circle). Thus the sought connection of the quantities is $v \cdot t = 2r\pi$. By dividing with t, we will express the unknown value v using the known values r and t: $v = \dfrac{2r\pi}{t}$. Now follows the insertion of the known values, and calculation:

$$v = \frac{2 \cdot 150\,000\,000\,\mathrm{km} \cdot 3.14}{1\mathrm{g}} = \frac{2 \cdot 3.14 \cdot 15 \cdot 10^7\,\mathrm{km}}{365 \cdot 24 \cdot 3600\,\mathrm{s}} =$$

$$\frac{2 \cdot 3.14 \cdot 15 \cdot 10^7\,\mathrm{km}}{3.65 \cdot 10^2 \cdot 2.4 \cdot 10 \cdot 3.6 \cdot 10^3\,\mathrm{s}} = \frac{94.2}{31.536} \cdot 10\,\frac{km}{s} \simeq 3 \cdot 10\,\frac{km}{s} = 30\,\frac{km}{s}$$

This is an extremely high speed. At this speed one would travel from London to New York in approximately 3 minutes. I leave it to you to remind yourselves about physics and consider why we do not feel this high speed of the Earth's rotation around the Sun. I consider it important now that you perceive how we calculated. The numbers were presented in an almost scientific notation. Then the small numbers were separately computed, the powers separately, and the units separately. This made the calculation clearer and simpler. □

Instruction for Practical Calculation

Calculation is done by presenting the numbers first in approximately scientific notation, as products of numbers close to the number one, powers of the number 10 and measurement units. Then, smaller numbers are computed separately, the powers of the number 10 separately, and the measurement units separately.

You can find more about scientific notation and computation in this notation at the site https://en.wikipedia.org/wiki/Scientific_notation

4.12 Coordinate System

We have introduced real numbers by means of geometrical considerations. Now we can reverse the perspective and express geometrical relations by means of real numbers. In particular, we can use them to fully describe the points on a straight line and relations between them.

Coordinate System on a Straight Line

By selection of a point on a line, a unit length and a direction (positive direction), through the measurement process each point is assigned a real number – the **coordinate of the point**. Here, different points are assigned different coordinates and every real number appears as a coordinate of a point. This assignment is called the **Cartesian coordinate system on a straight line**.

In order to see whether two points are equal, we look at their coordinates. In order to see whether one point is between another two, we look again at the coordinates.

Comparison of Points by Means of Coordinates

Points A and B are equal if their coordinates are equal.

Point A with the coordinate a is located between points B and C with the coordinates b and c precisely when its coordinate is between the coordinates of points B and C, that is if $b < a < c$ or $c < a < b$.

In the same way we could characterise also other geometrical relations of points on a straight line by means of coordinates. I leave it to you to describe, by means of relation *between*, what it means that two points are on the same side of a straight line in relation to a certain point on the straight line, and that you characterise this relation by means of coordinates.

Example 4.12.1.

1. Which of the points $A(-12)$, $B(-31)$ and $C(-10)$ is between the other two?

2. Which of the points $A(-4)$, $B(-7)$, $C(4)$ and $D(-1)$ are on that side of point $E(-3)$ where point $F(1)$ is also located?

Solution.

1. Since $-31 < -12 < -10$, point $A(-12)$ is between points $B(-31)$ and $C(-10)$.

2. Since $-7 < -4 < -3 < -1 < 1 < 4$, only points $D(-1)$ and $C(4)$ are on the required side.

$\square$

The coordinate connection between numbers and points can be used also in another direction, to represent relations among real numbers geometrically. For instance, we can simply represent the intervals of numbers:

> ### Intervals of Real Numbers
>
> **An interval of real numbers** is every non-empty set I of real numbers such that when it contains two numbers, it contains also all the numbers between them:
> If $a < c < b$ and $a, b \in I$ then $c \in I$.

The interval of numbers that contains all the numbers between a and b $(a < b)$ is designated $< a, b >$:

$$< a, b > = \{x \mid a < x < b\}$$

Such an interval is called an **open interval**. If we want to include the edges a and b in the interval, then we write:

$$[a, b] = \{x \mid a \leq x \leq b\}$$

Such an interval is called a **closed interval**. We can include only one edge in the interval:

$$\langle a,b] = \{x \mid a < x \le b\} \qquad\qquad [a,b\rangle = \{x \mid a \le x < b\}$$

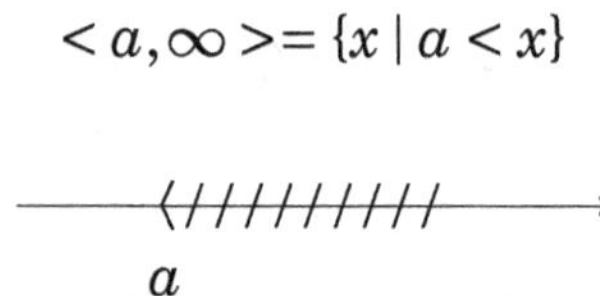

Intervals are also sets of all numbers bigger (or smaller) than the given number a (which may be included in the interval):

$$\langle a,\infty\rangle = \{x \mid a < x\}$$

(we read "interval from a to infinity")

$$\langle -\infty,a\rangle = \{x \mid x < a\}$$

(we read "interval from minus infinity to a")

The set of all numbers $\mathbb{R}$ is also an interval: $\mathbb{R} = \langle -\infty,\infty\rangle$

The coordinate system on a straight line is the base for introducing the coordinate system in a plane. The position of points in a plane can be measured in the following way. We will draw two mutually perpendicular straight lines. We will mark on each a coordinate system with the origin in the intersection of straight lines:

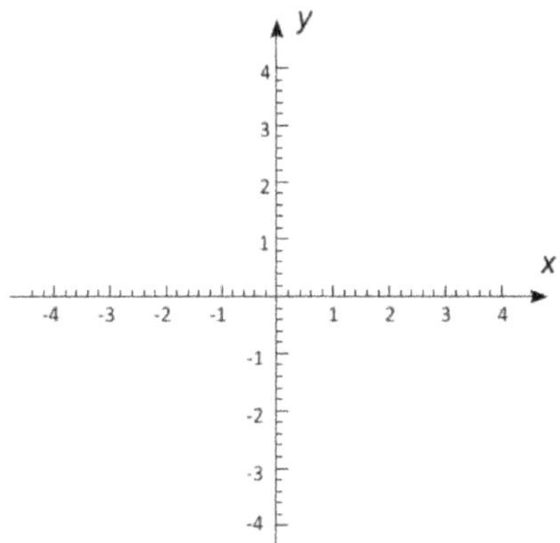

We will call the highlighted straight lines the x-**axis** and y-**axis**. Some also say **axis of abscissas** (lat. abscindere – intersect) and **axis of ordinates** (lat. ordinare – order) or **horizontal axis** and **vertical axis**. We will measure the position of the point T by means of its perpendicular projections onto the x-axis and y-axis. The projections are points Tx on the x-axis and Ty on the y-axis with the coordinates x and y:

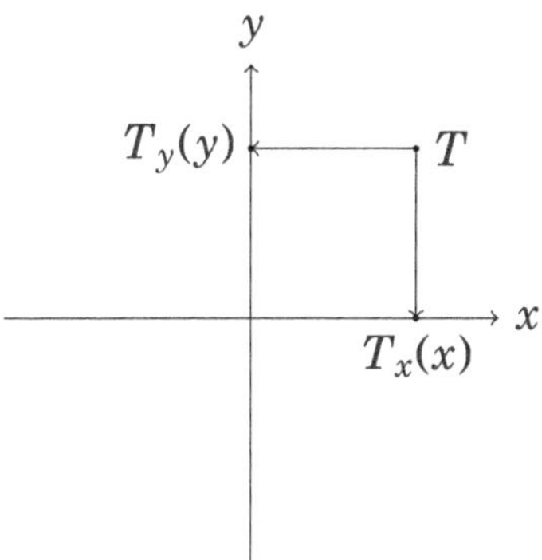

The coordinate x of point Tx will be called the **x-coordinate** (also **first coordinate** or **abscissa**) of point T, and the coordinate y of point Ty will be called the **y-coordinate** (also **second coordinate** or **ordinate**) of point T. We will see that these two numbers completely determine the position of point T.

Example 4.12.2. Let us determine the x- and y-coordinates of the following points:

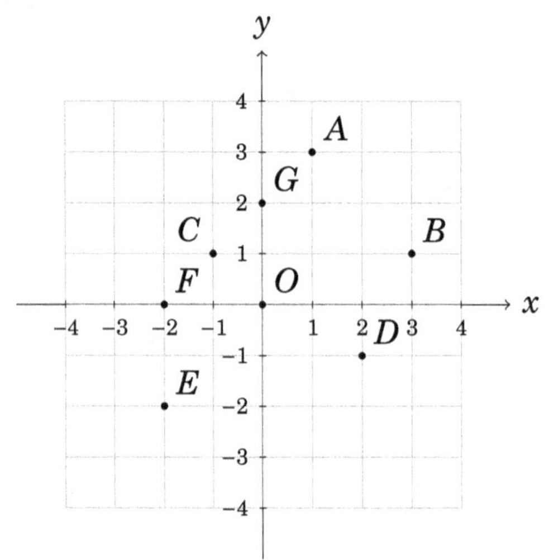

Solution.

	A	B	C	D	E	F	G	O
x	1	3	-1	2	-2	-2	0	0
y	3	1	1	-1	-2	0	2	0

It is not only important which numbers are coordinates of a point, but also which number is the first coordinate, and which is the second. In the previous example points A and B have assigned the same couple of numbers, 1 and 3, but the difference lies in the fact that point A has the first coordinate the number 1, and the second coordinate is the number 3, and in case of point B it is vice versa – the first coordinate is the number 3, and the second the number 1. With this procedure we do not assign a couple of numbers to a point but an **ordered pair** of numbers, a pair of numbers together with the information about which number in the pair is the first one, and which is the second.

Since different points have at least one different projection, different ordered pairs of coordinates are assigned to them. Therefore, the *ordered pair (x, y) of coordinates of point T can be considered as a measure of its position. Moreover, every ordered pair is the measure of the position of precisely one point.*

Example 4.12.3. Let us find the points with these coordinates:

1. $(1, 2)$ 2. $(1, -2)$ 3. $(1, 0)$ 4. $(0, 2)$

Solution.

1. The coordinates determine completely how to arrive at the point with these coordinates. The first coordinate shows how far and in which direction we have to travel along the x-axis, and the second one how far and in which direction we have to continue travelling parallel with the y-axis. First we arrive at position 1 on the x-axis, and then we move upwards by 2:

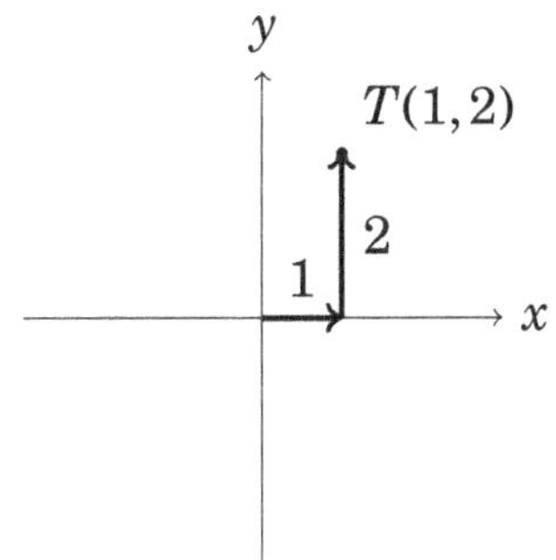

2. We also arrive at position 1 on x-axis, but we move downwards by 2:

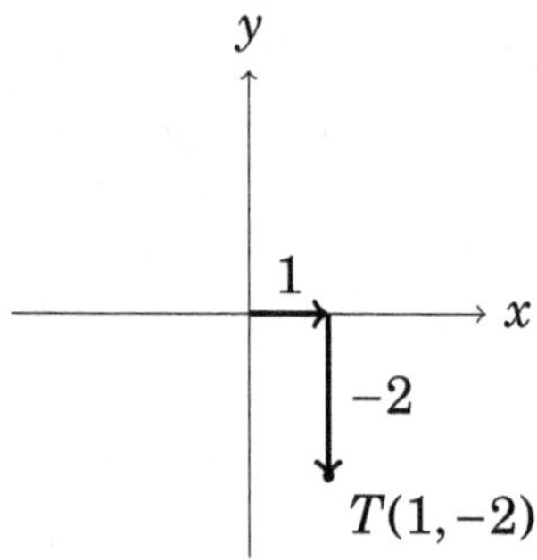

3. We again reach position 1, but we also stay there, and this is the sought point:

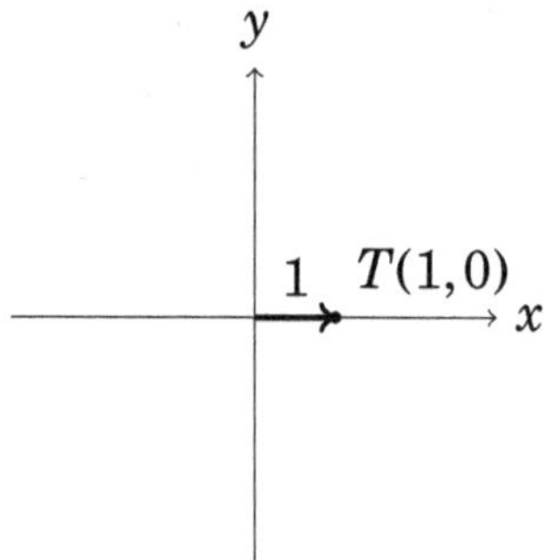

4. There is no horizontal movement now, but we move immediately upwards by 2:

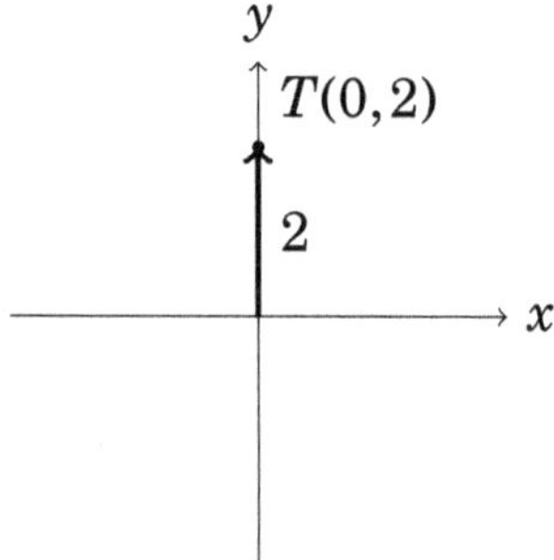

Coordinate System in a Plane

Through the described method, each point in the plane is assigned an ordered pair of numbers. Here, every ordered pair is assigned to only one point. This assignment is called the **Cartesian coordinate system in a plane** and it is completely determined by the choice of two mutually perpendicular straight lines, by the selection of a unit length and positive direction on each of these straight lines. Point T can be determined by means of its assigned pair (x, y), and every ordered pair (x, y) by its assigned point. This connection is often expressed by the notation $T(x, y)$.

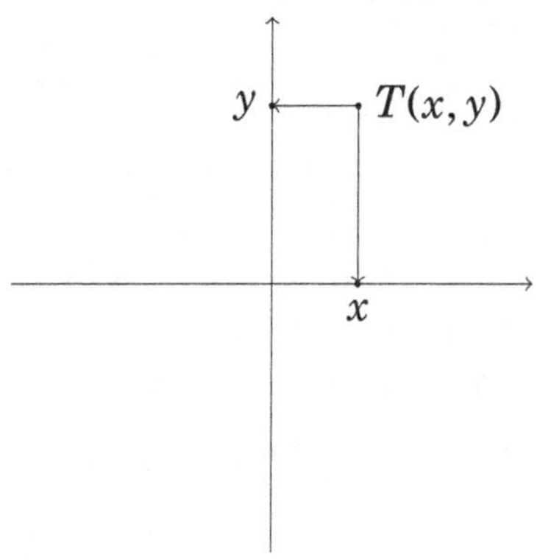

We do not have to say any more, "that point there", or "not that one but the one a little bit above", but instead we simply say "point A(2,3)", "point B(2.1,3.2)", and so on.

The Basic Idea of Analytic Geometry

A coordinate system connection of points and ordered pairs of numbers is the basis for translation from the language of geometry, which is pictorial, into the computationally efficient algebraic language, and vice versa. It was introduced by René Descartes in the 17th century and it proved very significant for the further development of mathematics. On the one hand, it facilitated solving geometry efficiently by the language of numbers – to compute geometry (such an approach to geometry is called **analytic geometry**); on the other hand, it enabled the visualisation of abstract objects from the world of numbers – to see algebra.

It is precisely this last aspect of "seeing" algebra that will be highlighted here, and in *Circle 2* it will be applied. Here, we will show how functions and equations with two unknowns are visualised.

René Descartes (1596–1650) has often been described as the father of modern Western philosophy, not because of his philosophical system but his method compressed into few words "Cogito ergo sum" ("I think, therefore I am"). The essence of his method is that the thinking of human beings themselves is the supreme authority of knowledge, and not the authority of church, state or public opinion. In the appendix of his book "Discourse on Method" (1637) he develops analytic geometry as an illustration of his method. As his philosophical work opened the way to modernity, this mathematical work opened the way to the expansion of European mathematics in 17th and 18th century.

The function (operation) f that assigns to a number x the number $y = f(x)$ is given by the open description which tells us what is to be done with x in order to obtain $f(x)$. Thus, for example, function f can be given by the following open description: $f(x) = x^2 - 2x$. Open descriptions are the only precise method of determining functions which map real numbers to real numbers. Knowing the open description, we can compute the function and test its properties. However, in situations where the description gets a little more complex, we do not "see" the properties of the function it describes. We do not see for instance, for which inputs we will get a positive value, and for which a negative value at output. Neither do we see for which input we will get the maximum value, and so on. It would be nice to have a presentation of the function in which all this is immediately visible. This method does exist. This is the graph of the function. It is the image of the function in the coordinate system from which we can read many items of information about the function.

Graph of a Function

A graph of a function f is a curve whose points have coordinates $(x, f(x))$.

In order to find for the given argument x the value of the function $f(x)$, we do the following:

1. we find x on the x-axis;

2. from this place we move vertically towards the graph;

3. when we reach the graph, we move horizontally towards the y-axis;

4. the coordinate of the place on the y-axis at which we arrive is the required value of the function $f(x)$.

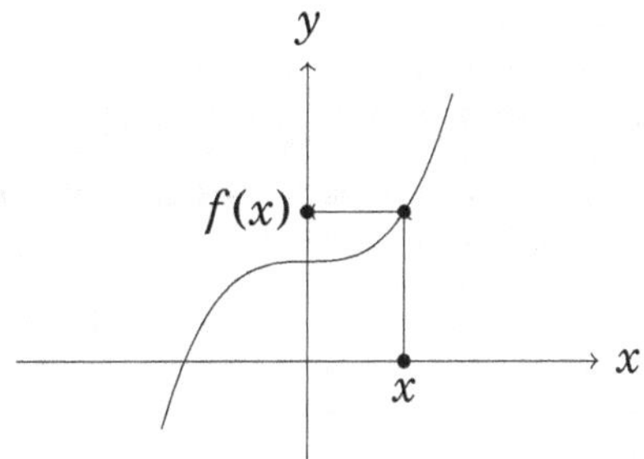

Example 4.12.4. For the function f given by the following graph let us determine $f(3)$, $f(-2)$, $f(0)$ and $f(-1)$.

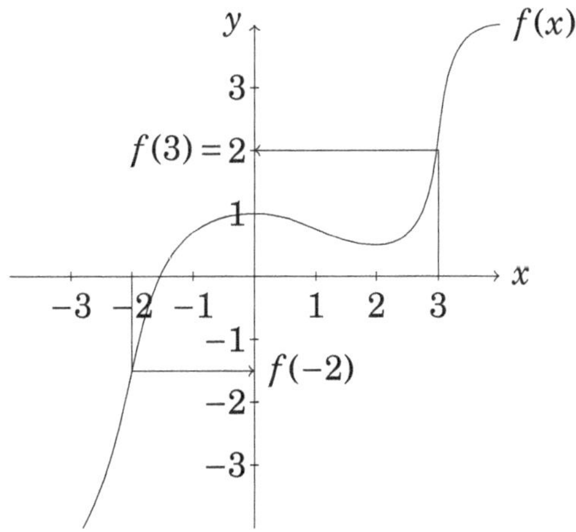

Solution. In order to find $f(3)$, from the number 3 on the x-axis we move upwards to the graph and then turn left towards the y-axis. On the y-axis we arrive at the number 2, so, $f(3) = 2$. In the same way we can determine also that $f(-2) = -1.5$; we only need to move from -2 on the x-axis downwards towards the graph, and then right towards the y-axis. In order to compute $f(0)$ we move from 0 on the x-axis towards the graph and when we reach it, we do not turn towards the y-axis because we are already on it, on the number 1. Thus, $f(0) = 1$. In computing $f(-1)$ we will not go from -1 on the x-axis towards the graph since we are already on it, but we will move immediately towards the y-axis at which we will arrive at the point with the coordinate 0. It is precisely such inputs in the function, which yield zero on the output, the so-called **zero points or zeroes of the function**, that will be important to us in *Circle 2* for solving equations. $\square$

In *Circle 2* we will see also how from the graph of a function we can qualitatively find out quite a lot about the function. Now we will only illustrate this by one example.

Example 4.12.5. When a stone is thrown vertically upwards at an initial speed of 10m/s it is known from physics that, if we ignore the air drag, its height y depends on the time passed t according to the following law: $y = 10t - 5t^2$. What maximum height will the stone reach, and when it will fall back to the ground?

Solution. I leave it to the reader to try, on the basis of their knowledge of mathematics, to answer these questions by examining the expression, which sets the dependence of the position on time. We will find here the answers by plotting the graph of the function using the software *SageMath*. The graph will be given by the command *plot*:

```
var('t')
plot(10*t-5*t^2,(t,0,3), ymin = 0, aspect_ratio=1)
# First input in the command is the function we are plotting,
# second input tells us the interval on which we plot it,
# the remaining two inputs are optional,
#- the first sets the lower limit to value plotting,
# since we are not interested in negative heights,
#- the second says that the same units are on both axes.
```

We will obtain the following graph:

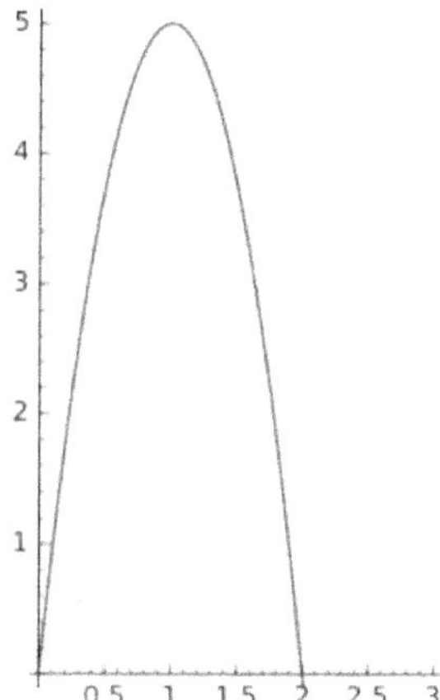

From the graph we see that the maximum height $y = 5$ has been achieved at moment $t = 1$, and that the stone will fall to the ground at the moment $t = 2$ (then the height is again $y = 0$). Also, from the graph symmetry we see that the falling occurs as if we had recorded the climbing and then reversed the film. In particular, the time of falling is equal to the time of climbing and takes 1 second. $\square$

When speaking about more complicated functions and more complicated questions, it becomes increasingly difficult to find the answers based on an open description which determines the function. On the other hand, from the function graph we can qualitatively read the answer, but also quantitatively up to a certain precision, in a way that is equally uncomplicated to when the functions are simple.

In *Circle* 2 we will see how the graph of a function helps us to solve qualitatively an equation with one unknown (to see how many solutions the equation has and to estimate them). The idea is simple. We transform an equation so that on the one side of the equation is zero. For example, we can write down the equation $x^3 = x$ (we haven't studied yet how to solve such equations) in the following form: $x^3 - x = 0$. The left side of the equation determines the function $f(x) = x^3 - x$. The solutions of the equation are just the inputs of the function which give zero at the output: $f(x) = 0$ (zero points of the function). And we can see them easily on the function graph: *zero points of a function are common points on the graph of the function and x-axes!* So, we will plot in *SageMath* the graph of the function $f(x) = x^3 - x$:

```
plot(x^3-x,(x,-2,2), ymin = -2, ymax = 2, aspect_ratio = 1)
```

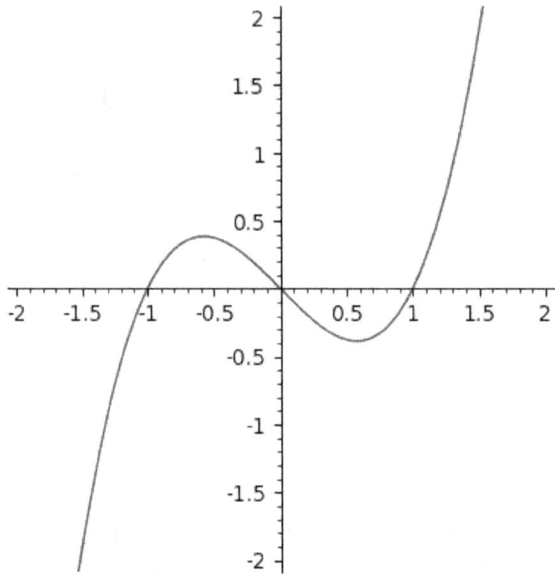

The common points of the graph and x-axes are 0 and 2 approximately, so they are the solutions of the equation (by inserting these values in the equation we can see that they are indeed the solutions).

The coordinate system allows us to present graphically also an equation with two unknowns. The equation with two unknowns is the condition on an ordered pair of numbers. The ordered pair of numbers can be understood as the coordinate of a point in the plane. Thus the set of all ordered pairs that are solutions of the equation can be presented by the set of points in the plane:

Graphs of Equations with Two Unknowns

A graph of equation $F(x, y) = 0$ is the set of all points $T(x, y)$ whose coordinates satisfy the equation.

With this connection the basic relation between an ordered pair of numbers and an equation with two unknowns (whether the pair of numbers is the solution of the equation) corresponds to the basic relationship between a point and a curve (whether the point belongs to the curve).

In the software *SageMath* we can plot the graph of an equation with two unknowns using the command *implicit_plot*. For instance, in order to obtain the graph of the equation $x^2 + y^2 = 1$ we will write

```
var('y')
implicit_plot(x^2+y^2==1, (x,-2,2),(y,-2,2), aspect_ratio = 1,
axes = true,frame = false)
# The first three inputs tell the software which equation
# we are plotting and in which area of the plane.
# Other inputs are optional - the last two define
# that we will get a coordinate system with axes and not frames,
# which is default for the command implicit_plot.
```

We will get a circle of radius 1 with its centre in the origin of the coordinate system:

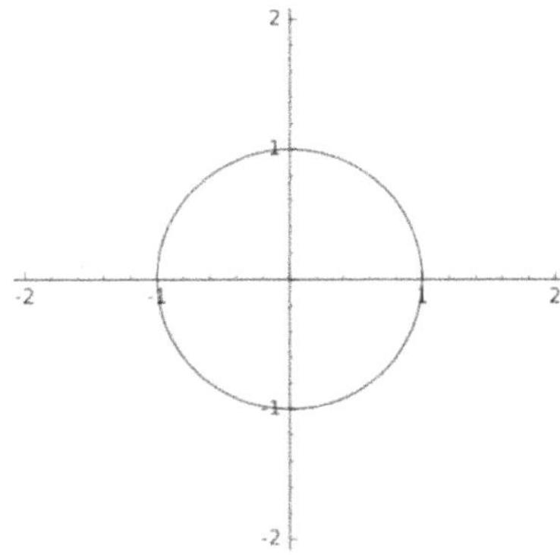

From the graph, we see that x can only be a number from interval $[-1, 1]$, and that for every value x within this interval there are exactly two values y that, paired with it, satisfy the equation, and they are of opposite signs. For marginal $x = -1$ and $x = 1$, the associated value for y is 0.

In *Circle* 2 we will see how graphs of equations help us to solve qualitatively a system of equations with two unknowns. Again the idea is simple. The solution of the system of two equations with two unknowns is an ordered pair of numbers which satisfies both equations. It means that the corresponding point in the coordinate system must lie in the graph of each equation. *So, looking at graphs of equations, the solutions of the system of equations are the coordinates of the common points of their graphs!* For example, we can find qualitatively the solutions of a complicated system of equations:

$$\left(x^2 + y^2\right)^2 = 4(x^2 - y^2), \quad x^2 + y^2 = 1$$

We will plot their graphs in *SageMath* (we will just add plot commands for each graph):

```
implicit_plot((x^2+y^2)^2==4*(x^2-y^2), (x,-2.5,2.5),
(y,-1,1), aspect_ratio = 1,   gridlines = 'minor')
+implicit_plot(x^2+y^2==1, (x,-2.5,2.5),(y,-1,1),
axes = true,frame = false, color = 'red')
```

We get:

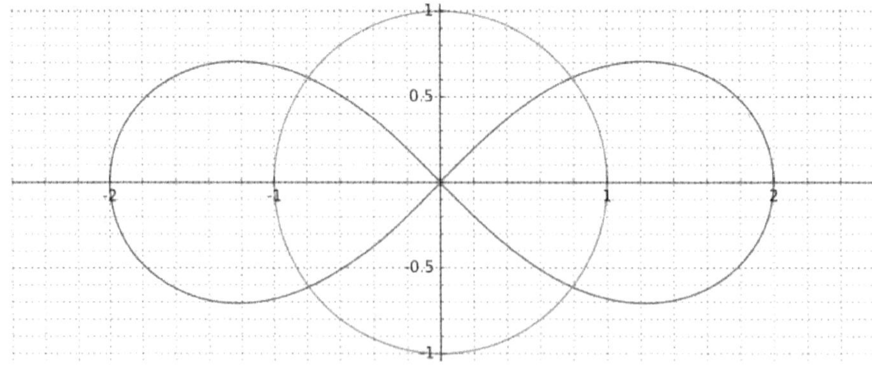

Now we can easily see the solutions (qualitatively): there are 4 mutually symmetric solutions which we can identify approximately: $(\pm 0.8, \pm 0.6)$.

4.13 On Mathematical Models

Have we solved the problem of measuring segments? The process of decimal measurement has been conceived so that, if necessary, we can always continue measuring, with units 10 times smaller. Is this feasible in real measurement? If not, does that mean that the theory of real numbers is actually unreal? In the end, what is the distance d from point O to point A that we started measuring at the beginning of this section? Measurement by units did not give any result:

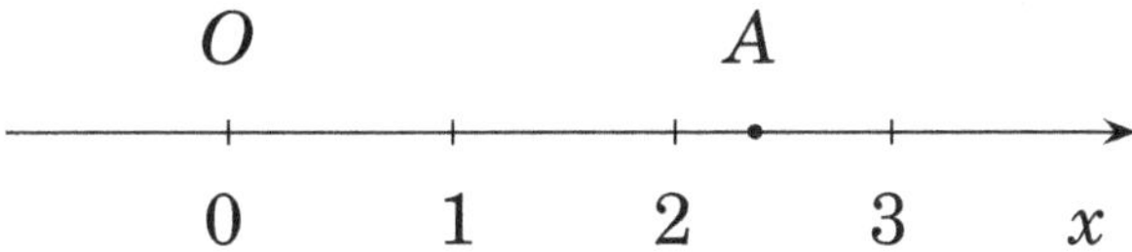

We obtained that $2 < d < 3$. The transition to tenths did not give a result either:

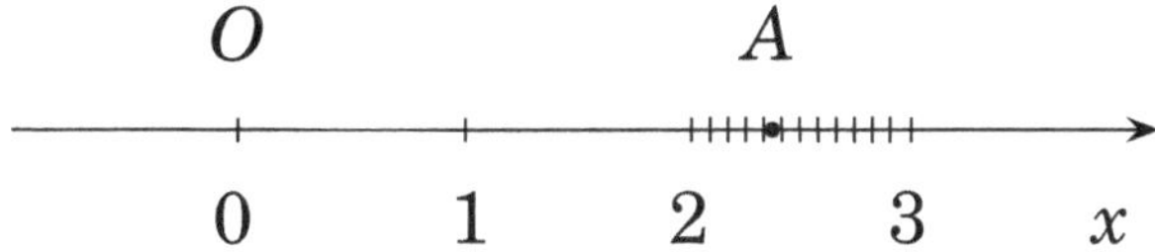

We obtained that $2.4 < d < 2.5$. For the majority of practical instances this is sufficiently precise and we can finish the measurement claiming that d is approximately 2.4. Finally, is there any sense at all in continuing to measure in hundredths or thousandths of centimetres or smaller? In reality, we cannot continue to measure by a ten-times-smaller unit ad infinitum, partly due to the available tools, and partly also because of what we are measuring. Namely, if we put what seems a point to us under a microscope, it will be a thick spot:

In the microscopic world this measurement problem proves to be incomplete. It seems that it should read: "How far is the left-est part of spot A from the right-est part of spot O?" However, neither would this formulation survive further enlargements. If we used a powerful electron microscope we would see a cluster of molecules in constant titration, so we would no longer even know what needed to be measured. And if we were to progress to an even smaller level, we would arrive into the area of quantum physics, where classical notions about objects and space are not valid. Not to mention yet smaller regions, about which we know even less.

The conclusion is that we have not solved the realistic measurement situation but rather an idealised situation of measurement, which we have conceived in a certain way. *We have replaced the real measurement problem by a simplified **mathematical model***. The mathematical model consists of an ideal straight line, ideal point and ideal process of measurement in which we can always continue to measure with a unit ten times smaller. And that is the problem that we solved with the concept of real numbers. But what happens with the real situation? Can our idealised model be applied to it?

> ### Application of Mathematical Models
>
> Mathematics gives precise models, but it cannot precisely describe the success of their applicability. This is a question that has to be continually repeated, and the answer is always open and uncertain.

We can only say that we know from experience that the previously described system of real numbers is applicable in all standard quantitative measurements. That is why it is so important. However, it may occur that in some future measurement it may prove unsuccessful. And what will we do then? We will try to find new concepts that will be successful in such situations.

As long as the laws of mathematics refer to reality, they are not safe; as long as they are safe, they do not refer to reality.

Albert Einstein (1879 — 1955)

4.14 Problems and Solutions

Decimal Notation

1. Convert the decimal notation into a fraction:

   (a) 1.3   (b) 0.0002   (c) −0.55

2. Convert the fraction into a decimal notation:

   (a) $\dfrac{131}{10}$ (b) $\dfrac{3}{20}$ (c) $\dfrac{77}{3}$

3. Compute without using a calculator:

   (a) $0.8 + 2.35$ (b) $3.19 - 4.293$ (c) $4.8 \cdot 5.6$ (d) $8.5 : 4.25$

Infinite Non-Periodical Decimal Notation

4. Determine which of the following numbers are rational and which are irrational:

   (a) 3.3222...   (b) −3.232232322323223...   (c) 4.45445445544555...
   (d) 6.7676...   (e) 9.345673456734567...   (f) −1.142143144145...

5. Which of the following statements are true?

   (a) Every integer is rational.

   (b) There exists a real number which is neither rational nor irrational.

   (c) All real numbers are rational.

   (d) Every rational number has a finite decimal notation.

   (e) Every finite decimal notation determines a rational number.

   (f) All irrational numbers have an infinite decimal notation.

   (g) There are rational numbers that have an infinite decimal notation.

Operations with Real Numbers

6. Compare the numbers $1.666766676667\ldots$ and $\dfrac{5}{3}$.

7. Find a sequence which gets closer and closer to the sum
   $3.14141414\ldots + 2.969696\ldots$.

8. Find a sequence which gets closer and closer to the difference
   $75.363636\ldots - 689.898989\ldots$.

9. Find a sequence which is closer and closer to the product
   $2.020202\ldots \cdot 1.101010\ldots$.

10. Find a sequence which is closer and closer to the quotient $\dfrac{8.080808\ldots}{2.020202\ldots}$.

Basic Properties of Real Numbers

11. Using geometrical reasoning prove the associativity of addition of the
    real numbers: $(a+b)+c = a+(b+c)$.

12. Using decimal notations and properties of rational numbers prove the
    commutativity of multiplication of real numbers $a \cdot b = b \cdot a$.

13. Prove both geometrically and using decimal notations that for $a > 0$
    and $b < c$, $ab < ac$.

14. Prove by the use of decimal notation that for every positive real number g there is a natural number which is greater than g.

Irrational Numbers

15. Among the following numbers find one rational and one irrational
    number:

    (a) $1.23442344\ldots$ and $1.23452345\ldots$ (b) $5.66666\ldots$ and 5.67

    (c) 1.4 and $\sqrt{2}$

16. Prove that the sum of a rational and an irrational number is an irrational number.

Approximate Calculation and Theory of Errors

17. For $a \simeq 0.023$ and $b \simeq 3.20$ determine how many significant digits there are in each number, and compute approximately

    (a) $a + b$ (b) $a - b$ (c) $a \cdot b$ (d) $a : b$

Units of Measurement

18. Convert $1.1\,\text{mm}$ into centimetres, decimetres and metres.

19. Convert $1\,\text{m}^2$ into cm^2, dm^2 and km^2.

20. A car is moving at a speed of 100 km/h. (a) How many metres will it travel in 1 hour? (b) How many kilometres will it travel in 1 minute? (c) How many metres will it travel in 15 minutes?

21. I claimed that the acceleration of free fall g is equal to 9.8 and my neighbour John asserted that it is approximately 32.1. If I referred to metres per second squared $\left(\dfrac{\text{m}}{\text{s}^2}\right)$ and he to feet per second squared $\left(\dfrac{\text{ft}}{\text{s}^2}\right)$ were we speaking about the same value for g ($1\,\text{m} = 3.28\,\text{ft}$)?

22. Domenico sent from America to Galileo (his brother who lives on a small Croatian island) 50 l of façade paint. The label on the can said that the coverage is $460\,\text{ft}^2/\text{gal}$ (square feet per gallon). If $1\text{m} = 3.28\text{ft}$ and $1\text{gal} = 3.785\,\text{l}$ how many square metres (m^2) will Galileo be able to paint with the paint he has obtained? The question is purely a theoretical one since Galileo has no intention of going to the trouble of painting the façade. In fact, Galileo claims that he has found the way to defeat capitalism in its essence. According to him, the essence of capitalism is to exploit people's work. Galileo has decided not to work, so capitalism cannot exploit him.

23. The shopkeeper Piero sold lasagne for 8433 lire, 2 soldi, 2 grossi and 20 picoli. If he paid 2820 lira, 4 solda, 3 grossa and 27 picoli for the pasta, how much did he earn? In the book *Treviso Aritmetica* (from the year 1478), which was written in the Venetian dialect of the Italian language by an anonymous writer, you will find, apart from similar problems, that 1 lira = 20 soldi, 1 soldo = 12 grossi, and 1 grosso = 32 picoli.

Scientific Notation

24. Convert the following numbers into scientific notation:

(a) $5\,987\,000\,000$ (b) 540 (c) 0.00751 (d) 0.00000051

25. Compute using scientific notation:

(a) $\dfrac{3600 \cdot 40\,000}{0.0012 \cdot 0.08}$ (b) $\dfrac{68\,000 \cdot 3000}{0.000034}$

(c) $\dfrac{(0.002)^2 \cdot 0.005}{200 \cdot (0.1)^3}$ (d) $\dfrac{(0.6)^2 \cdot 20\,000 \cdot 45}{300^4 \cdot (0.2)^3}$

26. The *NASA* telescope *Hubble* discovered a new galaxy unofficially called *Sharon* (officially STIS $123627 + 621755$). It has been estimated that light needed about 10^{10} years to travel from that galaxy to the telescope. How many kilometres is this galaxy away from us? The speed of light amounts to $c = 300\,000\,\text{km/s}$.

Coordinate System

27. Determine the coordinates of the following points:

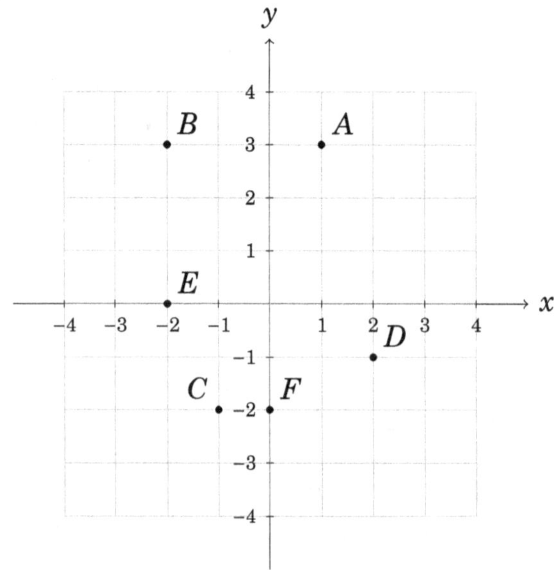

28. Find the points with the given coordinates:

    $A(3,5), B(-2,-6), C(5,-1), D(-3,4)$.

29. From the graph of the function $f(x) = x^2 - x^3$

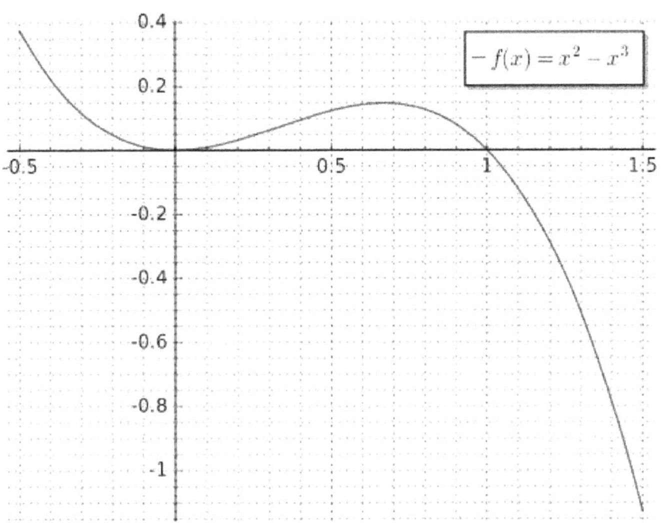

   (a) determine approximately $f(-0.5), f(0), f(0.5), f(1), f(1.5)$;

   (b) determine the sign of $f(0.8)$;

   (c) determine for which argument from the interval $< 0,1 >$ we will get the highest value of function and what value it is;

   (d) determine whether when we increase the argument from 0 to 0.5, the value of the function rises or falls.

30. From the graph of the equation $\left(x^2 + y^2\right)^2 = 4(x^2 - y^2)$

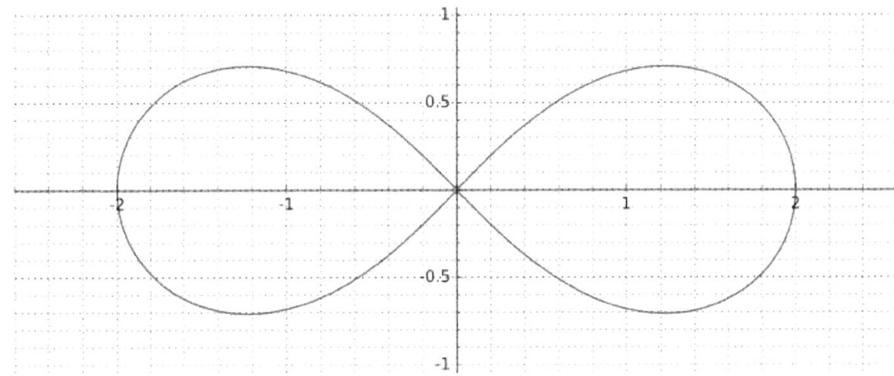

determine the solutions of the equation to which

(a) the first coordinate is 0.6   (b) the second coordinate is 0.5

(c) the first coordinate is 0

Additional Exercises

31. Mr. X started his car and drove for 8km, waiting at the traffic lights for 2 minutes. Mr. Y started his car and drove for 16 km without stopping. Using the table of data about the vehicles, compute who consumed more fuel and by how much

consumption	start	when idling	when driving
Mr. X	0.05 l	0.17 l per min	6 l per 100 km
Mr. Y	0.04 l	0.16 l per min	5 l per 100 km

32. The lawnmower consumes 0.6 gallons of petrol per hour. If we use it for 2.6 hours, how much petrol do we use? If 1 gallon equals 3.785 litres, what is the consumption in litres?

33. One small company produces toothpicks. The cost of the production of every toothpick packet is 0.5 kuna. Besides that, the company has a fixed expenditure of 5000 kuna weekly. If they produce x packets weekly, what is the average expenditure per packet? Due to pressure from bigger producers they sell every packet at a price of 1.5 kuna. Their work is profitable only if the profit on each packet is at least 0.5 kuna. If they make 2000 packets of toothpicks every week, is their business profitable or they will have to change something?

34. Scientists estimate that the Sun converts every second about $7 \cdot 10^{11}$ kg of hydrogen into helium. How much hydrogen does the Sun use during one year?

35. The mass of the Earth is approximately $5.98 \cdot 10^{24}$ kg, and the mass of the Sun is approximately $1.99 \cdot 10^{30}$ kg. How many times is the mass of the Earth smaller than the mass of the Sun?

Solutions

1. (a) $\dfrac{13}{10}$ (b) $\dfrac{1}{5000}$ (c) $-\dfrac{11}{20}$

2. (a) 13.1   (b) 0.15   (c) 25.666...

3. (a) 3.15   (b) -1.103 (c) 26.88   (d) 2

4. (a) rational   (b) rational   (c) irrational
   (d) rational   (e) rational   (f) irrational

5. (a) Yes   (b) No   (c) No   (d) No   (e) Yes   (f) Yes   (g) Yes

6. $1.6667666766667... > \dfrac{5}{3}$

7. $5, 6, 6.1, 6.11, 6.111, 6.1111, ...$

8. $-614, -614.5, -614.53, -614.535, -614.5353, ...$

9. $2, 2.2, 2.222, 2.22402, 2.2242402, 2.224260402, ...$

10. $4, 4, 4, 4, 4, ...$

11.

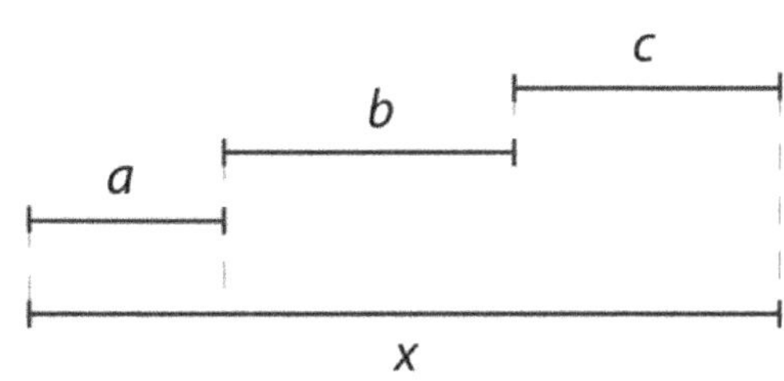

12. We will first look at the case $a, b > 0$. Let their decimal notations be
    $a = a_0.a_1 \dots a_n \dots$, $b = b_0.b_1 \dots b_n \dots$, Then

$$a \cdot b = (a_0.a_1 \dots a_n \cdot b_0.b_1 \dots b_n) \dots = (b_0.b_1 \dots b_n \cdot a_0.a_1 \dots a_n) \dots = b \cdot a$$

If either of the numbers a, b equal zero, then, analogously to the above procedure, we would show that $a \cdot b = 0 = b \cdot a$. If either of the numbers a, b is negative, commutativity can be proved by using the rules of sign and the previous result.

13. Here is the proof for $0 < b < c$.

    geometrically:

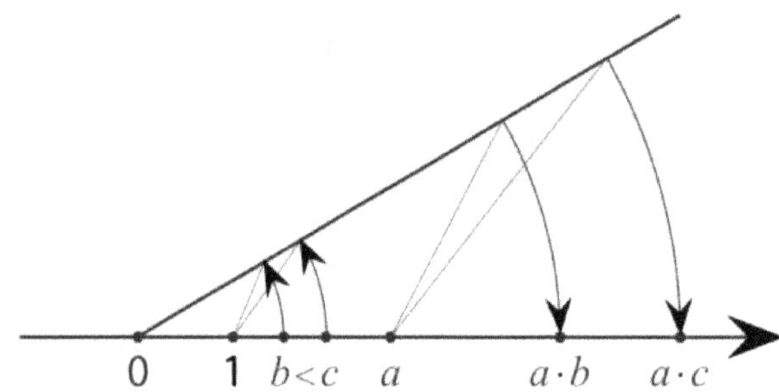

using decimal notation:

$b < c \rightarrow b_0.b_1 \dots b_n < c_0.c_1 \dots c_n$

$\rightarrow a_0.a_1 \dots a_n \cdot b_0.b_1 \dots b_n < a_0.a_1 \dots a_n \cdot c_0.c_1 \dots c_n \rightarrow ab < ac$

14. Let $g = g_0.g_1 g_2 \dots g_n \dots$. Per definition of comparison of real numbers $g_0 + 1$ is bigger.

15. (a) 1.2345 and 1.23452234222234...   (b) 5.667 and 5.6676667...
    (c) 1.401 and 1.401400140001...

16. Let i be an irrational and q a rational number. When their sum $r = i + q$ is a rational number, we would obtain that $i = r - q$ is also a rational number, which is a contradiction to the assumption that i is an irrational number. Thus, $r = i + q$ is an irrational number.

17. The number $a \simeq 0.023$ has two significant digits, and the number $b \simeq 3.20$ has three significant digits.

    (a) 3.22   (b) −3.18   (c) 0.074   (d) 0.0072

18. $1.1\,\text{mm} = 0.11\,\text{cm} = 0.011\,\text{dm} = 0.0011\,\text{m}$

19. $1\,\text{m}^2 = 10000\,\text{cm}^2 = 100\,\text{dm}^2 = 10^{-6}\,\text{km}^2$

20. (a) $100000\,\text{m}$, (b) $\dfrac{5}{3}\,\text{km}$, c) $25000\,\text{m}$

21. Yes

22. Approximately $565\,\text{m}^2$

23. We could convert everything into lira and then subtract. However, we can subtract also in a way that corresponds to subtraction in decimal notation:

24. (a) $5.987 \cdot 10^9$ (b) $5.4 \cdot 10^2$ (c) $7.51 \cdot 10^{-3}$ (d) $5.1 \cdot 10^{-7}$

25. (a) $1.5 \cdot 10^{12}$ (b) $6 \cdot 10^{12}$ (c) 10^{-7} (d) $5 \cdot 10^{-3}$

26. The path travelled is $s = c \cdot t$, where c is the speed of light, and $t = 10^{10}$ years. Thus,

$$s = 300000\,\frac{\text{km}}{\text{s}} \cdot 10^{10}\,\text{years} = 3 \cdot 10^5\,\frac{\text{km}}{\text{s}} \cdot 10^{10} \cdot 3.65 \cdot 10^2 \cdot 2.4 \cdot 10 \cdot 3.6 \cdot 10^3\,\text{s}$$

$$= 3 \cdot 3.65 \cdot 2.4 \cdot 3.6 \cdot 10^5 \cdot 10^{10} \cdot 10^2 \cdot 10 \cdot 10^3 \cdot \frac{\text{km}}{\text{s}} \cdot s = 94.61 \cdot 10^{21}\,\text{km}$$

27. $A(1,3), B(-2,3), C(-1,-2), D(2,-1), E(-2,0), F(0,-2)$

28.

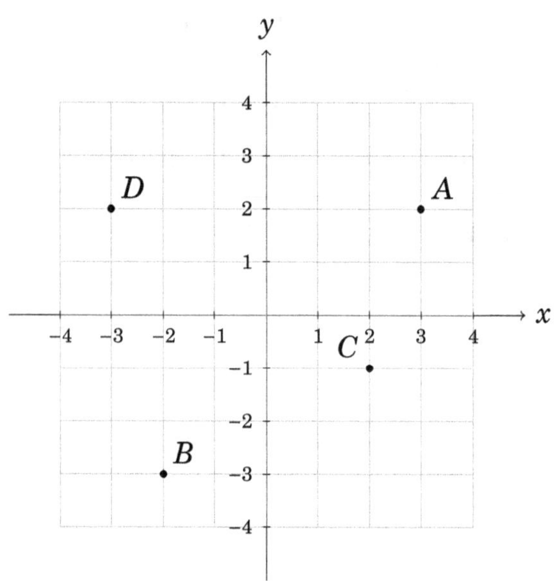

29. (a) $0.37, 0, 0.13, 0, -1.15$

    (b) the sign is positive

    (c) For the argument 0.66 we will get the maximum value of function 0.15.

    (d) rises

30. (a) $(0.6, 0.5), (0.6, -0.5)$ (b) $(0.6, 0.5), (-0.6, 0.5), (1.8, 0.5), (-1.8, 0.5)$

    (c) $(0, -2), (0, 0), (0, 2)$

31. Mr. X consumed 0.031 petrol more than Mr. Y.

32. $1.56\,\text{gallons} \simeq 5.9\,\text{litres}$

33. The average cost per packet is $0.5 + \dfrac{5000}{x}$ kuna. For $x = 2000$ the average cost is 3 kuna per packet. Since per packet they make an income of 1.5 kuna, not only do they not make a profit, but they are losing 1.5 kuna. Thus, they need to change something urgently.

34. $2.207\,52 \cdot 10^{19}\,\text{kg}$

35. $3.327\,76 \cdot 10^5$ times

Inspiring End

We have realized the concept of measuring in various ways, using various systems of numbers, depending on what was measured. For every kind of measurement we have conceived an idealised measuring process and an adequate system of numbers as results of the respective measurements. Thus, we used natural numbers to measure how many items are in a certain set, integers to measure how many oriented items are in a certain set, rational numbers to measure the oriented parts resulting from the division of a whole into equal parts, and real numbers to measure arbitrary parts of arbitrary oriented items. However, the concept of measurement has not been exhausted in this way. We have seen how with the introduction of a coordinate system we can measure the points in a plane by pairs of numbers. In the same way we can measure points in space by ordered triplets of numbers.

Using ordered quadruplets we can measure events – to the space coordinates of the place at which the event has occurred we have to add also its time coordinate. For instance, the event described in the figure has the coordinates (1,3,1.8,17). This means that it occurred at the place (1,3,1.8) (a well-known Zagreb meeting point *under the clock* on the main square) at 5 p.m.

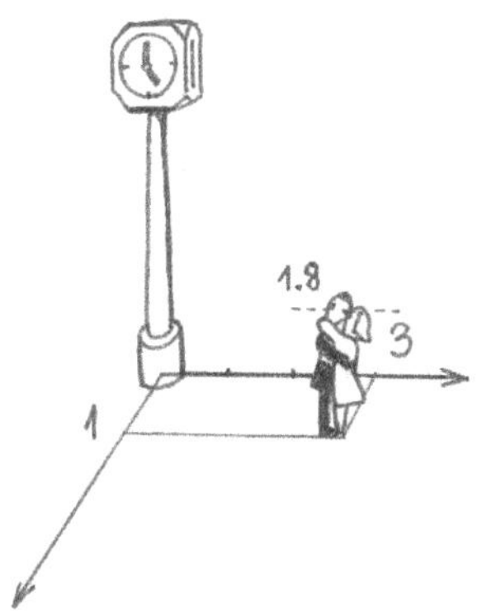

Using ordered quintuplets we can describe, for example, the status of your material assets. A quintuplet (45,18,4,0,1) may mean that you own 45 CDs, 18 books, 4 balls, 0 kuna and one pet. The state of your computer display is measured by a whole table of numbers, where every number measures the state of one point of the display. The measurement of the state of temperature in your apartment is a function which assigns to every place in your apartment the temperature at that place. In *Circle 3* we will go on to measure the relations between points using vectors, alternate currents will be measured by complex numbers, geometrical transformations by matrices, symmetries by the so-called symmetry groups, and so on. In the mid-19th century, starting from the idea that geometry is not a science of space itself but rather about the method of measuring it, the mathematician Bernhard Riemann formulated the general mathematical concepts of measuring not only space but also various other sets in which objects are continuously interconnected. These so-called (Riemann's) **manifolds** became half a century later the mathematical base of the general theory of relativity, and they are today one of the basic concepts in mathematics and modern physics.

Various measures have been developed in mathematics. For some of them it holds that various items have the same measure. For instance, we used rational numbers to measure the parts of gold bars. Each pirate got his share of gold, but (pirate) justice was satisfied since the parts had an equal measure. For some measurements, however, it holds that different items have different measures. For instance, different points in a plane are assigned different ordered pairs of numbers. It is said then, that the measures **represent** the measured values, that they are their representatives. The goal is to use the representatives to facilitate testing of certain properties. In this way, for example, geometrical properties of points are reduced to the calculation with coordinates that represent them. The world of numbers is so rich that it can represent every other world. Since the world of real items is usually finite, we can represent them with natural numbers. To various items we assign different natural numbers.

There need even not be any deeper relation of the item and its representative number. We also speak about **coding** (Lat. codex – piece of wood for writing) of items by numbers. Thus, every person in Croatia is assigned a PIN, a unique identification number of the citizens that identifies them in the state system.

Coding is sometimes used for the transfer of secret messages. Messages encoded by numbers can be read only by the person who knows the decoding rule. Therefore, the goal is to set the least possible transparent coding. Such coding is called **ciphering** (Arab. sifr – digit). Without ciphers (and mathematics in the background) it is almost impossible to conceive the modern world.

However, successful coding is usually based on a completely opposite principle, and that is to describe the relations among the items by simple calculation relations among their codes. The conceptual base of the computer contains precisely such coding. The data are coded by a series of zeroes and ones. Their processing is thus reduced to calculation with zeroes and ones. This is precisely what lies in the background of your favourite computer game, as well as in the background of the creation of this text that I am right now (with my last powers) typing into the computer, using LaTeX, a text processing software. However, I cannot resist mentioning one more miracle that coding made possible. In the 1930s, a young Austrian called Kurt Gödel had the idea of coding mathematics itself. He used natural numbers to code the mathematics of natural numbers and he discovered things that shook the foundations of very logic, mathematics and other sciences, and due to which many philosophers and mathematicians cannot sleep peacefully even today. We shall speak about this in more detail in *Circle 4*. Briefly, he showed that there are limits in describing our conceptions – we can neither describe them fully, nor prove that they are consistent. This holds for all even slightly more complex conceptions, not only for mathematics but also for physics, computer science, philosophy, law,.... He discovered that for any statements about a certain more complex conception we use for axioms, there will always exist statements about the conception that we will not be able to prove nor refute from these axioms. Our conceptions remain always an open story that needs to be supplemented all the time. Also, he discovered that we cannot prove that our somewhat more complex conception is consistent (free of contradictions), except if we use a more complex conception whose consistency is even more questionable. These results have destroyed the dream of some about mathematics as a firm base and eternal truth. I see it differently. It is precisely this imperfection and uncertainty that render mathematics such an exciting human product. I would personally far rather participate in this than in some boring eternal truth. I do not consider this as a sign of its weakness but rather as a sign of its life. Mathematics is a human creation and it shares the characteristics of all the other human creations as well as our lives.

I used to measure the skies, now I measure the shadows of Earth.

(Part of the epitaph on the grave of Johannes Kepler,
an astronomer and a mathematician, 1571–1630)

Calm Epilogue

In *Circle 1* we did not only study numbers; rather, we used the world of numbers to illustrate what mathematical objects are, how we describe them and how we apply them. Let us repeat the basics.

In what ways mathematical objects exist is a question that even today has not been completely solved. Fortunately, mathematics functions very well even without an answer to this question. My attitude is close to Dedekind's: mathematical worlds and mathematical objects in them do not really exist, but are rather imagined worlds with the purpose of being an effective tool in our understanding of the real world. For instance, numbers are conceived as the results of an idealised process of measuring that enables us to describe more precisely the world and its laws. Not only can we use measurements to be able to define something more precisely, but also regularity in nature can be precisely expressed by the relations between numbers. An important characteristic of all good mathematical concepts is that they are an effective tool for studying and shaping the world. Such a view of mathematical objects has a certain advantage, since it gives us the freedom to imagine any type of world, not caring whether it exists or not. We need not have any natural-scientific or metaphysical or religious and mystical cognitions about the world in order to understand the mathematical worlds and to know what the eternal truth is about them, and what is not. These are simply our ideas, that we can eventually also modify if this may seem necessary to us for any reason. Even historically, such a conception of mathematics has proved to be significant, so significant that we can regard it as the basis of modern mathematics.

An important part of mathematics is **mathematical language**. Many who think that they do not understand mathematics in fact do not understand mathematical language, its form and meaning. It is very important to have a good notation of objects in which the content-wise relationships (the relationships between described objects) transfer into simple formal relationships (relations between linguistic descriptions of these objects). This allows us a clearer and more efficient management, particularly in more complicated situations. Moreover, the formal part can be handed over to the computer to do for us. Let us remember only the decimal notation of numbers. It is the same with sentences which describe mathematical worlds. Good sentence form allows simple and precise expression and deduction, which can be partly left for the computers to do. Let us remember only the equations and simple procedures for solving them. An essential part of the effectiveness of mathematical language results from the use of symbols and variables. Symbols allow us to make simple calculations even with the most complicated descriptions and statements. The variables, the names of intentionally unspecified objects, allow us a very significant leap in thinking and calculating from the concrete (about single objects) to the abstract (about any objects). Using variables, we infer general laws about all objects in as simple a way as we infer also particular truths about concrete objects.

Another important part of mathematics is **proofs**. From the truth of one statement we prove the truth of other statements. This renders mathematics greater precision, safety and efficiency and uncovers its logical structure. Usually, the imagined mathematical world is described by one set of basic assertions that are called the axioms of that mathematical world. By proving from the axioms, we discover in the safest way possible what is valid in that mathematical world and what is not. The truth of all the statements is reduced to the truth of the axioms and correct thinking. In this book we have described real numbers axiomatically. Thus, our intuition about numbers has assumed a completely precise form without any doubts as to what is and what is not. Deduction from the axiom is especially important in complicated situations and in situations in which we have no developed intuition. Mathematics is usually developed as a combination of intuition and axiomatic deduction. Both components are important and inseparable within the body of mathematics. The axiomatic method improves greatly our economy of thinking because every statement that we deduce from a set of axioms is valid in every mathematical world which obeys the axioms. In this way we also discover similarities between mathematical worlds.

Yet another important part of mathematics is **definitions**. A precise description of a mathematical world requires not only highlighting of the basic truths of the world based on which all the others are proved, but also of the basic symbols, words and concepts which compose the description and whose meaning we understand. If we want to use a new symbol or word, we have to define it using the old ones. This is done using a definition: the description which we use to specify the meaning of a new symbol using the meanings of the old symbols. Definitions reduce the meaning of every word to the meaning of the basic words and correctness of definitions. This renders the mathematical story clearer – everything that is said has a precise meaning and no ambiguity. Moreover, by replacing (by means of definitions) entire descriptions with a single word, we simplify the expression and structure the thinking. Definitions are one of the basic mechanisms of abstraction that enable us to successfully deal with the complexity of content.

No matter what objects are considered, there is naturally the need to consider **sets** of these objects, **relations** between them and **operations** (**functions**) that assign to one object other objects. Since they occur in all the considerations, they are the basic mathematical objects. As such, they are not just auxiliary mathematical tools but they are themselves also a subject of mathematical consideration.

Common to all mathematical thinking is also the very action of **thinking**, so that also is a subject of mathematical thinking. Although we have a developed sense for correct thinking that we correct by exercising and knowing what correct thinking is, we can still think more correctly; moreover, we can hand over a part of thinking to the computer as well.

An important criterion of the validity of a particular mathematical world is not its truthfulness but rather its **applicability**: whether it is successful, directly or via other mathematical worlds, in our cognition of the world. Numbers are the oldest and still the most successful mathematical concept. We have seen also with the example of numbers, that mathematics gives precise models but cannot precisely describe the success of their application. Also, via the example of numbers we have seen that mathematical concepts, although they are often calculated formally through appropriate notation, are applied content-wise, according to their meaning. Therefore, it is important to understand the mathematical concepts. Without understanding them, they are inapplicable. Unlike formal procedures, understanding is a human activity in which computers cannot replace us. With the example of equations, we have illustrated a typical application of mathematics. Knowing the laws of the field from which the problem has originated, and with some simplifications, we have translated the actual problem by means of the meaning of mathematical concepts from natural language into mathematical language. The translation is a mathematical object (in our case that was the system of equations). We also say that we have obtained a mathematical model of the problem. In mathematical language we usually develop efficient formal methods that yield a solution. However, they yield the solution to a mathematical problem. This solution should be interpreted to see whether it is the solution of the actual problem. It is possible that the translation from the actual problem into the mathematical problem was bad. However, the art of translation is not only the matter of knowing mathematics but also of knowing the area in which the problem has arisen. Furthermore, in the example of equalities which express laws of nature we have illustrated another important typical application of mathematics – it provides us with efficient tools for inferring unknown laws of nature from already known ones.

In this first Circle we have dealt with numbers, which served as an example of mathematical objects. In considering numbers, we have developed various mathematical tools and met various characteristics of mathematics. We will refine and expand all that in the Circles to come.

Index

www.ingramcontent.com/pod-product-compliance
Lightning Source LLC
Chambersburg PA
CBHW060111120726
48003CB00009B/2591